THE INFINITE BOOK

John D. Barrow is Professor of Mathematical Sciences in the Department of Applied Mathematics and Theoretical Physics at the University of Cambridge. He is the author of several best-selling books, including *Theories of Everything*, *Impossibility*, and *The Book of Nothing*.

ALSO BY JOHN D. BARROW

Theories of Everything
The Left Hand of Creation (with Joseph Silk)
L'Homme et le Cosmos (with Frank J. Tipler)
The Anthropic Cosmological Principle (with Frank J. Tipler)
The World Within the World
The Artful Universe
Pi in the Sky
Perchè il mondo è matematico?
Impossibility
The Origin of the Universe
Between Inner Space and Outer Space
The Universe that Discovered Itself
The Book of Nothing
The Constants of Nature: From Alpha to Omega

John D. Barrow

THE INFINITE BOOK

A Short Guide to the Boundless, Timeless and Endless

VINTAGE

Published by Vintage 2005

2 4 6 8 10 9 7 5 3 1

First published in Great Britain in 2005 by
Jonathan Cape

Vintage
Random House, 20 Vauxhall Bridge Road,
London SW1V 2SA

Random House Australia (Pty) Limited
20 Alfred Street, Milsons Point, Sydney,
New South Wales 2061, Australia

Random House New Zealand Limited
18 Poland Road, Glenfield,
Auckland 10, New Zealand

Random House (Pty) Limited
Isle of Houghton, Corner Boundary Road
& Carse O'Gowrie, Houghton 2198,
South Africa

The Random House Group Limited Reg. No. 954009
www.randomhouse.co.uk/vintage

A CIP catalogue record for this book
is available from the British Library

ISBN 0 099 44372 4

Papers used by Random House are natural, recyclable products made from wood grown in sustainable forests. The manufacturing processes conform to the environmental regulations of the country of origin

Typeset in Bembo by
Palimpsest Book Production Limited, Polmont, Stirlingshire
Printed and bound in Great Britain by
Cox & Wyman Ltd, Reading, Berkshire

To Luca Ronconi
for his boundless imagination

'A lot of English writers are consigned to writing about adultery in Camden because the big themes are elsewhere'

Damon Galgut[1]

Contents

Preface

'I am painting the infinite'

Vincent van Gogh[2]

This is a book about the biggest subject of all. It's the ultimate traveller's guide to everything that could be: the rough and the smooth guide to infinity.

Infinity has haunted human minds for thousands of years. It challenges theologians and scientists alike to understand it, cut it down to size, find out if it comes in different shapes and sizes, and decide whether we want to outlaw or welcome it into our human descriptions of the Universe. Is it part of the problem or part of the solution?

It is also a live issue. Physicists' accelerating quest for a Theory of Everything has been primarily guided by an attitude towards infinities. Their appearance can be a warning that you have entered a blind alley on the road to the truth. The enthusiasm with which superstring theories were embraced was a consequence of their ingenious evasion of the problem of infinities that had plagued all their predecessors.

These exciting new theories leave us to decide whether we should expect matter to be infinitely divisible. Will we always be able to find ever smaller, more elementary, particles inside any that we have, like a never-ending sequence of Russian dolls? Or is there a limit, a smallest 'thing', a smallest size, or a shortest time, where division comes to a full stop? Or perhaps the fundamental entities out of which the world is woven are not really little particles at all?

Cosmologists have their own problems with infinities. For decades they have been happy to live with the notion that the Universe of space and time began at a 'singularity', where its temperature, its density, and just about everything else, was infinite. But will the marriage of gravity and the quantum really permit actual infinities? Is their appearance a sign of success or failure? Are infinities just a signal that we have not found enough pieces of the puzzle, or are they a vital part of the solution to ultimate problems like the beginning and end of the Universe, the moments of the Big Bang and the Big Crunch?

Cosmologists have another strange infinity to contemplate: the possibility of an infinite future. Does the Universe seem to be on course to last forever? What does 'forever' mean? Can life in any form continue forever? And, at a more human level, what would it mean – socially, personally, mentally, legally, materially, and psychologically – for us to live forever?

Mathematicians have also had to face up to the reality of infinity. The issue was a big one, one of the biggest that mathematicians have ever faced. Just seventy years ago, mathematics faced a civil war over the meaning of infinities, leaving many a casualty and much bitterness. Some wished to outlaw infinities from mathematics and redefine its boundaries to exclude all treatments of infinities as real 'things'. Journals were closed down and mathematicians ostracised because of their attempts to exclude infinities from mathematics.

At the root of all the fuss was one man's work. The genius of Georg Cantor showed how to make sense of the paradoxes of infinity that Galileo had first identified three hundred years before. What is the nature of an infinite collection? How can it be that you can take things away from it and it still stays infinite? Can one infinity be bigger than another? Is there an ultimate infinity beyond which nothing bigger can be constructed or conceived, or do infinities go on forever? But Cantor didn't live long enough to see the fruits of his genius form part of the acknowledged body of mathematics. Sidelined and undermined by influential opponents of infinite mathematics, he gave up mathematics for long periods, was encouraged when his ideas were enthusiastically taken up by Catholic theologians, yet suffered from long bouts of depression and illness before dying alone in a sanatorium. One of the neglected heroes of mathematics, a talented artist, a simple genius: one of our chapters will tell his moving story.

Theologians ancient and modern have struggled to make sense of the infinities lurking within their doctrines and beliefs. Is God infinite? Must he not be 'bigger' than other more mundane infinities, like the never-ending list of all positive numbers? What do different religions make of infinities? Are they regarded as a threat or a suggestion of something transhuman? Cantor provides a completely unexpected answer.

Ancient philosophers, beginning with Zeno, were challenged by the paradoxes of infinities on many fronts, but what about philosophers today? What sort of problems do they worry about? We will give some examples of live issues on the interface between science and philosophy that are concerned with whether it is possible to perform an infinite number of tasks in a finite time. Could a real computer perform a super-task? What would happen if it did? Of course, this simple question, in the hands of philosophers, needs some clarification: like what exactly is meant by 'possible', by 'tasks', by 'infinite', by 'number', by 'finite', and, by no means least, by 'time'.

As we range more widely through modern science we encounter an array of strange problems about infinity: is the Universe finite or infinite? Will it go on forever? Is the past infinite? Can *anything* happen in an infinite Universe? Are there problems that would take an infinite time for any computer to solve? What are those problems like?

Most people think of infinity and boundlessness as one and the same thing. Curiously, they are not. There are finite things, like the surface of a snooker ball, that have no boundary at all. A fly could walk around it forever without encountering an edge. Curved spaces are different – but what happens if they become infinitely curved? And didn't Einstein show us that outer space is curved, so what does this tell us about the Universe?

There are also unusual ways in which time can be finite yet not have an end. Usually, we think of time as a straight line stretching out in front of us. Time seems straightforward. Every event is either in the future or in the past of any other event. Alas, the Universe is not so simple. Take a straight line of soldiers marching one behind the other: each of them can say who is in front of them and who is behind. But make them march in a circle and now everyone is both in front and behind everyone else! There is no ordering any more. If time becomes circular in an analogous way, it allows time travel to occur and all manner of strange paradoxes can be conceived. You read this book carefully and travel backwards in time to tell me, word for word, all that's in it. So where did the idea for this book come from? You got it from me, but I got it from you. It seems to have been created out of nothing – a bit like the Universe.

I would like to thank Luca Ronconi, Sergio Escobar, Pino Donghi, Bruna Tortorella, Serafino Amato, Guilio Giorello, Paul Davies, Michael Brooks, Jörg Hensgen, Will Sulkin, Gary Gibbons, Joseph Dauben, Carl Freytag, Janna Levin, Stephen Clark, and Steven Brams for their help and input at various times as this book came to fruition. It is dedicated to Luca Ronconi with special thanks for his enthusiastic creativity which made *Infinities* such a success in the Italian theatre. I would also like to thank Elizabeth for her infinite patience with this project and our children, now no longer children, who remain unconvinced that we need any more books and thought that this one sounded as if it might be a long one.

chapter one

Much Ado about Everything

'On a clear day you can see forever.'

Alan Lerner[1]

THE ROUGH GUIDE TO INFINITY

'If there is a Universal and Supreme Conscience I am an idea in it. After I have died God will go on remembering me, and to be remembered by God, to have my consciousness sustained by the Supreme Conscience, is not that, perhaps, to be immortal?

Miguel de Unamuno[2]

There is something about infinity and books. Never-ending stories, libraries that contain all possible books, books that contain everything that has ever happened, and everything that hasn't; books that write themselves, books about themselves, books about there being no books, and books that end before they've begun. So you should be no more surprised to find yourself reading a book about infinity than I am to be writing one. But for something that you can't buy on the internet, 'infinity' is strangely ubiquitous. It turns up in church sermons, mathematics lectures at all the best universities, popular science books about 'Life, the Universe and Everything', and mysticism the world over, while historians remind us that people have been burnt at the stake for talking about it. It is at once the staple of the mystic contemplation of reality – 'make me one with everything' as the mystic said to the hamburger vendor – and the familiar territory of science fiction and fantasy. Can all these things really be connected? Is infinity really that big?

For thousands of years in the West there was no more seditious idea on Earth than that of infinity. The idea that things might go on and on forever, that they need have neither beginning nor end, neither centre nor boundary, was contrary to the wisdom of the West. It threatened to displace God Almighty from His uniquely infinite status, to demote the Earth from the centre of the Universe, and destroy the uniqueness and special meaning of every event in creation. It had the potential to make what was once merely the possible become inevitable.

Yet the temptation to think that way was strong and simple. Once you start doing something over and over again it's not too hard to imagine what it would be like never to stop. Infinity is just one thing after another. And this tantalising mixture of simplicity and sophistication remains with us today. Infinity is a subtle idea to capture precisely and easy to throw into the dustbin of wishful thinking, but for the ordinary person in the street it is less surprising and more readily intelligible than any comparable abstraction. We are immune to its subtleties; protected by a strange familiarity inbred by religious traditions, or from just staring out at the dark night sky; convinced by our method of counting that there could never be a biggest number. If in doubt just add one. Or can you?

Yet infinity remains a fascinating subject. It lies at the heart of all sorts of fundamental human questions. Can you live forever? Will the Universe have an end? Did it have a beginning? Does the Universe have an 'edge' or is it simply unbounded in size? Although it is easy to think about lists of numbers or sequences of clock 'ticks' that go on forever, there are other sorts of infinity that seem to be more challenging. What about an infinite temperature or an infinite brightness – can such physical things actually be infinite? Or is infinity just a shorthand for 'finite but awfully big'? These sorts of infinity seem more problematic than the unending futures promised to the followers of many traditional religious faiths. Eternal life doesn't need anything infinite to happen here and now. It just means that there will always be something happening – always a there and then.

The other religiously motivated infinity is that which goes loosely with the idea of a God of limitless power and knowledge, which is a key ingredient of many Western religious traditions. This is another familiar touchstone for the concept of the infinite for everyone. You don't need to be a mathematician to feel that this type of transcendental infinity is familiar. Or do you?

You do need to be something of a mathematician to appreciate the other type of infinity. Numbers go on and on. Infinity seems to be nothing more than where they would get to if counting went on forever. But surely it never does and mathematical infinity looks like a promise that is never fulfilled, a numerical Peter Pan, a shorthand for a goal that is never reached, a potential but not an actual, a number bigger than all numbers. Or is it?

Already we begin to sense that there are different sorts of infinity and you might believe in one but not another. In this book we are going to explore these infinities from different directions. We will see how human thinking came to embrace the idea of the infinite before recoiling from its implications. We will see how the argument raged about whether any true infinity ever materialised in our finite Universe; or whether infinities were artefacts of an inadequate description of events, are invariably relegated to happen in the infinite future, or are excluded from reality by a hidden principle that upholds the logical consistency of the Universe. We will find that eventually mathematicians became accustomed to dealing with infinities as if they were real entities, adding and subtracting them, cataloguing all the different infinities, determining their sizes, and finding that some were bigger than others – infinitely bigger. But we will mingle our story with tales that make the paradoxes of the infinite grow to become as large as life.

INTIMATIONS OF THE INFINITE

'think globally but act locally'

Activist bumper sticker'[3]

We know where the famous 'lazy eight' ∞ symbol for infinity came from. The Oxford mathematician John Wallis, who was famous for writing the codes for both sides in the English Civil War, first wrote down the symbol in 1655. With a few strokes of his pen he adapted the Roman representation ⊂|⊃ sometimes used instead of M for the (for them, large) number 1000. When written quickly it became ∞ and it stuck. This and other uses of this evocative symbol can be seen in Figure I.I.

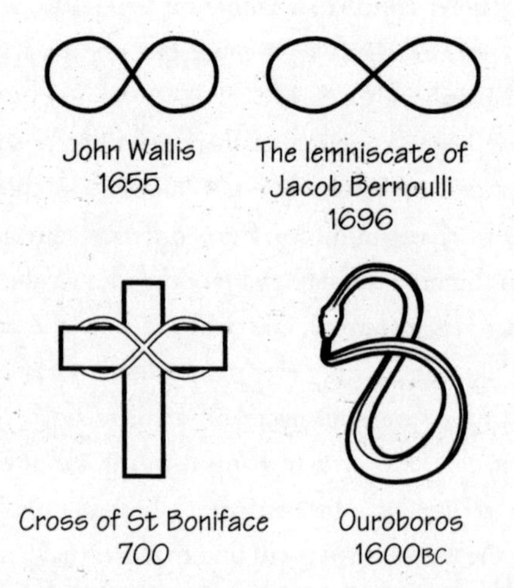

Fig 1.1 *Some examples of the 'lazy eight' symbol. John Wallis used it as a symbol for mathematical infinity in 1655. Jacob Bernoulli drew the curve of the Lemniscate[4] in 1696. The distinctive Cross of Saint Boniface appeared around 700 while the ancient symbol of the Ouroboros – the snake eating its tail – can be found as early as 1600 BC.*

Where did the idea of infinity come from? Does it bring with it some subtle survival value that favoured those with the inclination to develop it? Evolutionary psychologists would look for some way of thinking or acting which aided survival on African savannah landscapes a million years ago and had as a by-product the liking for generalisation without end. Nothing specific is immediately obvious. Primitive life was brief and immediate. Action was needed. Contemplation was not rewarded. The inclination to think about infinity is something that happens much later in the human story and it emerges from one of many responses to the Universe around us. What are the trails that might lead to forever?

There is a single pattern to many of the intuitions that have led human minds to contemplate the infinite. Human consciousness enables us to look ahead and see patterns. This enables us to compress experience into formulas or symbols that are shorter than the experience itself. We can write histories. This compressibility and pattern in the world is what ultimately makes mathematics so useful to us: we can pick out the patterns that are evident and represent them by strings of numbers or symbols. These strings generally have the property that they require no end. A list can always be added to. They naturally give credence to the idea of sequences of events that go on forever, even if there is no physical evidence that they do.

The idea that time has no end

> 'Eternity's a terrible thought. I mean, where's it all going to end?'
>
> Tom Stoppard[5]

'Immortality', it has been said, 'is the bravest gesture of our humanity towards the unknown.'[6] This is not an obvious response to the nature of everyday reality. Human beings, like other living things, are mortal. You would need to be a philosopher to distinguish clearly between time and our experience of it. The easier thought is to notice that time goes on for us when others die. The seasons may come and go, but there is a constant cycle of growth and decay and regrowth. The psychological responses to this state of affairs were various. For some, the response to human mortality was to regard it as an illusion or an antechamber to a more complete form of existence which was endless. The completeness of this higher form of existence was defined by its never-ending quality. For others, human life-cycles were like those of other living things and we would be reborn as part of a cycle of changes. Both of these ideas lead to an expectation of endless existence by extrapolating from what we see around us to create a satisfying perspective on the Universe in which we occupy a meaningful place. Ideas like these can play an important role in binding groups of people together, maintaining their morale in the face of adversity, and inspiring them to give their lives in defence of their fellows.

The idea that time has an end is at least as hard to maintain as the belief that it doesn't. What would it mean? What would it feel like? It only made sense if there was some great cataclysm in the future that would destroy everything – but even in mythologies where such a drama was played out, something always happened next. Bringing time to an end seemed to involve having no actors, no gods to determine the fate of the world. Strangely, in the Christian world we have grown up with the naturalness of a world with a beginning and an end and do not worry about the mind-stretching problems of a world with neither beginning nor end – that just always is. But it is surely the finite world that seems strangest. It needs someone or something on the outside to bring it about in order to provide it with a context and a reason to be. Take away our religious heritage and it may have been more natural to assume that earthly things go on without end. But, paradoxically, it is our Christian religious heritage that reinforces an expectation that things go on forever, with or without us . . .

'World without end.
Amen'

Cycles

'Like a circle in a spiral
Like a wheel within a wheel
Never ending or beginning,
On an ever spinning wheel
As the images unwind
Like the circles that you find
In the windmills of your mind.'

Alan Bergman and Michel Jean Legrand, *Windmills of Your Mind*[7]

In many cultures there was a strong belief that all change is cyclic. There is good reason to think so. Everyday life witnesses to it. Birth, life and death lead to rebirth; night follows day as day follows night and the seasons recur with metronomic regularity. Sleeping and waking, our lives are a continually repeating cycle. What better place to look for a picture of the ultimate pulse of the Universe?

Some believed in a more specific form of cyclicity in which all living things were reincarnated on Earth in the guise of other creatures. Other religions believed in rebirth by transformation into a new body and soul. In essence all these religious ideas of resurrection and rebirth look to a future without end but with change. Like a ball bouncing forever, so they look to a future that has no end and draw from a past that had no beginning. Invariably, human beings had a part to play in that never-ending cycle of existence. Life is a process, a flow, in which we emerge temporarily but are subsumed and replaced by other living things. A beginning or an end would be a singularity, a disruption of the natural order of things. Such a hiatus would be unnatural, inexplicable without the invention of other forces at work in the Universe. Psychologically, having a place in an infinite process endows the believing participant with a part to play in the infinite scheme of things, a sense of community with all living things, and a personal trajectory that is ever renewed.

The Supreme Being

> 'God is more truly imagined than expressed, and He exists more truly than He is imagined.'
>
> St Augustine[8]

Many cultures had a conception of a Supreme Being who controlled the Universe. In most cases this Being was the first among many, the leader of the gods. In others he was unique in certain respects, all powerful and all knowing. If such a Deity controls everything, even space and time, He cannot be limited by them and so must be eternal or transcend time entirely. Again, we see how one is led to entertain an idea of what we would call the infinite. It is a necessary attribute of a certain type of Deity.

This type of search for the infinite is also closely linked to a human desire for something transcendent, something beyond what is seen and immediately experienced. Some would argue that this inclination arises because there *is* something transcending our immediate experience. This is the stance of the great religious traditions. Others argue that this is a by-product of the unusual development of the human mind. At some stage in our evolution our minds developed an ability for self-reflection. This enabled us to imagine what would happen if we took certain actions. This is a remarkable ability. Other animals don't seem to have it. They learn by direct experience rather than by imagined experience. This type of human consciousness has all sorts of by-products, and creates fears and psychological problems from which simpler minds will not suffer. Is our tendency to extrapolate from the known to the unknown and on to the unknowable a by-product of the mind's ceaseless attempts to correlate what we know?

Unending space

'The eternal silence of these infinite spaces terrifies me'

Blaise Pascal, *Pensées*

The greatest shared experience of human beings throughout their history has been the appearance of the night sky.[9] The darkness of the night sky, studded with bright celestial objects, was a remarkable feature of ancient life. It inspired stories, provided the means to navigate, and elicited worship. It gave humanity a sense of place in the greater Universe – and that place was a humble one. We appeared as an insignificant dot amidst the star-spangled blackness of the night. That blackness went on and on, perhaps forever. How could it end? Again, the idea of a cosmic edge is harder to grasp than that of its absence. What world lies beyond such an edge and where would it be? The dark night sky might be a great dark shell that surrounds us, like a celestial cave wall – with lights upon its ceiling. Or if you live on an island or a continent that is partially bordered by the sea, you will have seen that there can exist a complete change of environment. There could be an edge to space in the way that there is an edge to land at the coast. What lies beyond need not be nothingness, merely something different, something that we choose not to call space.

In the book of Genesis there is one of the greatest of stories, a heroic quest by men to reach for the sky and be like gods. The Tower of Babel[10] is presented in the end by the authors of Genesis as a tale to explain the origin of human languages. Yet, this was far from what its builders intended as they planned their ziggurat 'whose top may reach unto heaven'.[11] Ever since, Babel has been a by-word for the hopeless quest, tilting at windmills, the Millennium Dome of the ancient world.

For some, the thought of unending star-spangled space is a lure, pulling humanity out of itself towards something superhuman and beyond; but, more often than not, that cosmic fascination inspires feelings of fear as much as awe. The seventeenth-century French philosopher Blaise Pascal told us of his awe and terror in the face of the potential infinity of space around us and the fleeting span of his life in the eternity of time, while the Russian artist Wassily Kandinsky talked of the 'great silence, like a cold indestructible wall going on into the infinite'.[12]

Today, we understand much more about the vastness of space and the nature of what is 'out there'. There is still a pull towards discovering the unknown, but there is also something quite different. We are the first generations of humans who have been able to see the Earth from space. The special appearance of the Earth amidst the vastness of space played an important role in accelerating the appeal of environmental movements in the 1960s. Our very finiteness in the centre of immensities made us pause to rethink the direction in which human technology was taking us and the risks that it was storing up for the finely balanced environment that makes Earth both an inhabitable and a beautiful planet.

Counting

> **'The animals went in two by two,**
> **Hurrah! Hurrah!'**
>
> Traditional children's song

The study of ancient systems of counting and reckoning tells us a lot about what it was they liked to count as well as how they did it.[13] The simplest systems began with number words for one and two which they could combine in simple ways to count larger quantities. Often the words for these low numbers were specific to the objects being counted. There would be different words for two stones and for two hands. This is something we still see today in English where there are many special words for two, depending on what is being enumerated: a brace, a duet, a pair, a double, a twosome, a couple, a doublet, and so on. Only a few ancient cultures took special interest in very large numbers and the simpler cultures were content to describe any large number by a word that might express the idea of 'many' either literally or graphically, say by alluding to the hairs on your head or the sand grains on the sea shore.

If you wish to do your sum on stone, or clay, or paper, then you soon become concerned with notation: the search for a succinct way to record numbers. The system we use today is beautifully simple. It was inherited from ancient India. It allows us to represent any finite quantity by using only the symbols 0, 1, 2, 3, 4, 5, 6, 7, 8, and 9. The secret is that the relative positions of these symbols carry information about their meaning. Thus, whereas for a Roman centurion III would mean the number three, for us it means one hundred and eleven (100+10+1). There is no end to the lengths of the strings of symbols we can create with the Indian decimal system.

However, there was another way of thinking about numbers without having a system of recording them. The most primitive intuition is to add or take away one from what you have got. If you put a stone on the ground every time one of your sheep enters the field in the morning and then take a stone away as each one passes back out at sundown, you can tell whether any sheep have been lost without ever counting them. If you start adding to your pile of counters, one by one, you begin to see that there is nothing to stop you adding one more. All you need is another counter. To keep on counting without stopping, you need to have a never-ending supply of counters. How you respond to this state of affairs determines whether you conclude that you can count forever. If you think the physical counters are essential, then they will run out unless the world contains an unlimited number of counters. But if you believe the latter, you already have the notion of the infinite. If you don't think the counters are essential, then there is never anything to stop you adding one. It's like the children's game of the biggest number. If you choose a number, I can always find a bigger number by adding one. Yet, to implement this strategy in practice, you need to be able to express the bigger number in words and it must be evident that it is a bigger number. In our decimal system we can always make a number that is ten times bigger than the one we started with simply by adding a zero on the right-hand end of the number you have given me; so, for example, 34 grows to 340.

The never-ending list of numbers shows the vital role that has been played by the notations and symbols we have devised to represent quantities. The Indo-Arabic numeral system we employ is wonderfully economical and suggestive. It can be used to record any number, however large, given a big enough piece of paper. It does not suddenly fail when numbers reach a critical level. It does not require the invention of a new symbol. As a result it positively encourages an intuition that numbers go on forever. However, as we shall see, that is not quite the same thing as infinity.

Subdivision

> 'He watched her for a long time and she knew that he was watching her and he knew that she knew he was watching her, and he knew that she knew that he knew; in a kind of regression of image that you get when two mirrors face each other and the images go on and on and on in some kind of infinity.'
>
> Robert Pirsig, *Lila*[14]

When we hear mention of infinity we tend instinctively to think big – stars, galaxies, and the incomprehensible vastness of unending space – but there is an infinite inner space waiting in the palm of your hand. Keep cutting something in half and the fragments get smaller and smaller. How far can you go? Can you keep cutting them in half forever, or is there ultimately a smallest piece, an indivisible building block at the bedrock of things? The ancient Greeks were more enamoured of the puzzles that were created by trying to make infinitely many subdivisions of space and time than of trying to imagine the infinitely large. Indeed, we see that the idea of endless subdivision is a more manageable and familiar one than that of the ready-made concept of a limitless universe or of a space that has no edge. One can always see practical barriers to cutting things in half – you always need a smaller knife or a shorter wavelength of laser light – but it is easy enough to imagineer your way around them. You also readily see that no singular end point of zero size is ever reached. An infinitely large universe *is* actually infinite, but the infinitely small is always more easily seen as potentially so if it is imagined as the culmination of a cutting process that in reality never ends. In the words of William Blake, we may hope

> 'To see a World in a Grain of Sand,
> And a Heaven in a Wild Flower,
> Hold Infinity in the palm of your hand
> And Eternity in an hour.'[15]

Subdivision is special because it has a practical dimension. There are advantages to making small things and thin surfaces – narrow blades and pieces of miniature design. As you do this you begin to realise that there is a practical barrier to making thinking smaller. This barrier arrives rather more quickly that you might expect. Take an ordinary piece of A4 paper and try folding it in half more than seven times. You won't be able to do it. Halving moves quickly and soon the thickness of the paper becomes the same size as its diameter. Instead, try cutting the paper in half time after time. You'll get to about twenty-two cuts and then start to struggle. Eventually the size of the cutting instrument starts to be a limiting factor. So it must have been for any process of breaking rocks or wood. Subdivision brings you face to face with the limitations of the actual, no matter how much you might think about a process that could go on potentially forever.

Fig 1.2 *Some ancient Islamic tilings.*[16]

Patterns

> 'The number you have dialled is complex. Please rotate your phone through ninety degrees and redial.'
>
> Answer-phone message

One of the most persistent suggestions of the infinite in finite human affairs has been our universal aesthetic sense. All human cultures have displayed a desire to create art and music. Empty spaces are a provocation to make patterns and designs, to create images and patterns that inspire, instruct and illuminate. In some cultures there have been religious vetoes on the representation of living things which channelled the creative impetus into a fascinating exploration of the infinite in finite form (Figure 1.2). The most impressive ancient examples are to be found in the Islamic world where the tessellation of flat and curved

spaces explored all of the mathematical symmetries that we now know to be possible.

These designs suggest the infinite in two ways. First, explicitly by the recipe for indefinite repetition. The repeating of the same design, side by side, over and over again is an algorithm that reaches out to infinity. It needs no boundary to complete it. It was once thought that if you wanted to tile a flat surface with tiles of one or two shapes, then the resulting pattern would have to be periodic, repeating just the same over and over again. The simplest example – that you have probably used on a floor or a wall somewhere in your own house – is just to use a single square tile; but you can be a bit more adventurous and use equilateral triangles or even hexagons (Figure I.3).

However, not all regular polygons will do the same trick. You will not be able to tile your bathroom floor with regular 5-sided tiles. No matter how you shift them around there will always be some unfilled surfaces.

In 1974 Roger Penrose discovered the remarkable fact that two tile shapes can be used to cover a flat surface of infinite extent without simple periodic repetitions. One of Penrose's tiling pairs is called the dart and kite and is shown in Figure I.4.

Second, and more subtly, we now appreciate how to build up patterns by means of repetition that changes its scale at each application. These

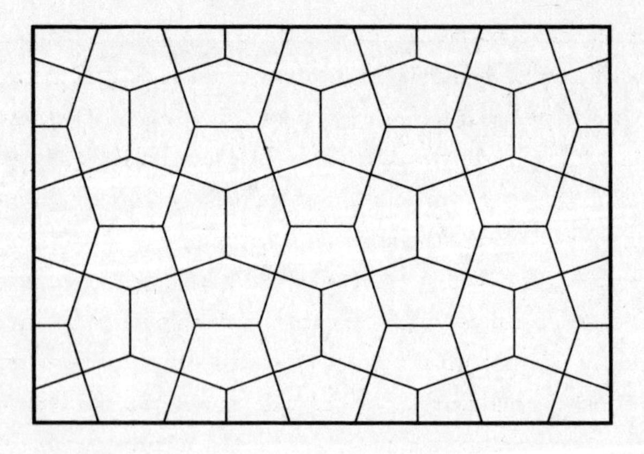

Fig 1.3 *Periodic tilings of a flat surface by pentagons which combine in force to form hexagons.*

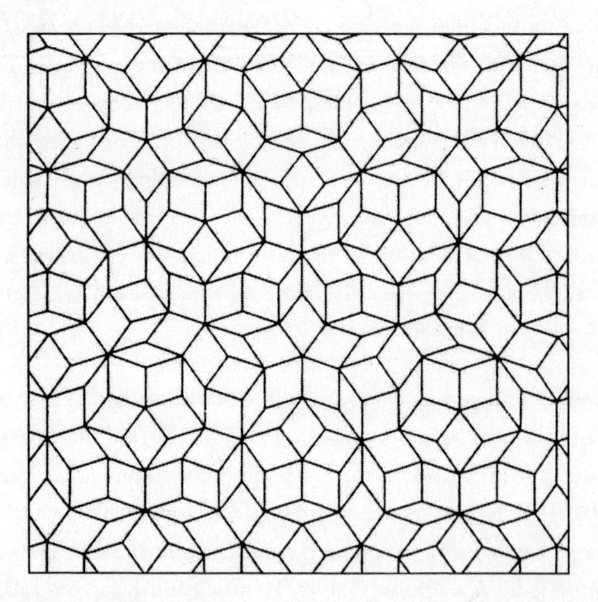

Fig 1.4 *Roger Penrose's infinite but non-periodic tiling of the plane using darts and kites.*[17]

designs are called fractals and they can be found all over the natural world, in the patterns of branching trees, plants, and clouds. Anywhere, in fact, where a volume needs to be enclosed by a very large surface. By making the surface wiggly and crenellated it is possible to enlarge the surface and increase the capacity to cool, to absorb nutrients, or to interact with other things without a proportionate increase in volume or weight. In Nature the repetitive process that creates a fractal never repeats infinitely often, but it creates a process of pattern creation that points to infinity. As an example, we can see what happens if we make an equal-sided triangular notch from the middle third of each side of an equal-sided triangle and then keep on repeating this process. Curiously, it turns out that the area that is enclosed by the zig-zag boundary line is always finite, while the length of the boundary line becomes as long as one likes. Again, the creation of a design process points to a never-ending sequence of steps which has a final outcome that can easily be imagined.

There is an ancient curiosity about infinite patterns that is as striking today as when it was first appreciated nearly 2500 years ago. Suppose that we think about the world of two dimensions that exists flat on the surface of this page. Now ask how many possible figures, 'polygons', there are with numbers of equal straight sides that can be drawn on it. Start with three sides and there is a triangle, then with four sides – a square, then the five-sided pentagon, and so it goes on *forever.* We have drawn some of them in Figure 1.5. Notice that as the number of sides gets bigger, so the figure looks more and more like a circle.[18]

It is clear from these pictures that there is no end to the number of flat polygons that can be drawn with numbers of equal sides. But now

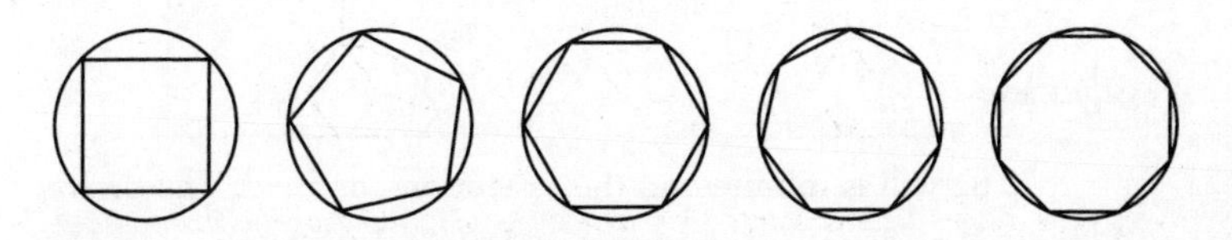

Fig 1.5 *Some of the regular polygons. There are an infinite number of varieties and they can have any number of sides, exceeding two.*

let's ask the same question in three dimensions. How many solids are there with flat sides of equal area? The simplest of these polyhedra, with four faces, each triangular, is a tetrahedron. Next comes the cube, with six faces. Then comes the surprise, first discovered long ago by the early Greeks: there are only three more, the octahedron (eight faces), the dodecahedron (twelve faces), and the icosahedron (twenty faces). The names come from the number of faces and they are all shown in Figure 1.6.

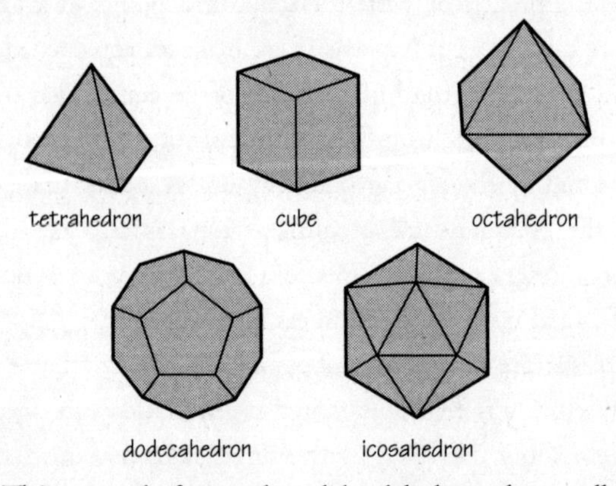

Fig 1.6 *There are only five regular solid polyhedra – the so called 'Platonic solids'.*

These five, the only possible regular solids, are known as the Platonic solids. What is so striking is that stepping up from flatland into three dimensions, seemingly into a space where there is much more 'room', turns out to be hugely constraining. What is an infinite collection in two dimensions becomes small and finite in three dimensions.

Possibilities

> 'The will is infinite and the execution confined. The desire is boundless and the act is a slave to limit.'
>
> William Shakespeare, *Troilus and Cressida*

If you have the virtue of consciousness, so that you can think about the future and not merely let it happen to you, then intimations of the infinite can creep up on you in another way. Freewill is a funny thing. We can't help thinking we have it. We seem to be able to think what we like. There is no obvious limit on what we can think and the imaginations of our minds. They may not be remarkable; they may not be useful; but they seem to be always slightly different. New experiences, new contexts and new interactions create a continuous spectrum of different impressions and pictures of the world. This I believe is an important motivation to think that there are limitless possibilities and an infinite store of possibilities that we can dip into. Of course, despite feelings to the contrary, there are only a finite number of thoughts that we can have. The number is huge – the biggest number you have probably ever seen – but it is none the less finite. By counting the number of neural configurations that the human brain can accommodate, it has been estimated that it can represent about $10^{70,000,000,000,000}$ possible 'thoughts' – for comparison there are only about 10^{80} atoms in the entire visible Universe.[19] The brain is rather small, it contains only about 10^{27} atoms, but the feeling of limitless thinking that we possess derives not from this number alone but from the vastness of the numbers of possible connections that can exist between groups of atoms. This is what we mean by complexity, and it is the complexity of our minds that gives rise to that feeling that we are at the centre of unbounded immensities. We should not be surprised. Were our mind significantly simpler, then we would be too simple to know it.[20]

ZENO HOUR

'There was an old lady who swallowed a fly'

Nursery rhyme

While there are subtle intimations of the infinite in the things we do and see, there are also deep paradoxes that lie very close to the surface of things. The most ancient and the most famous of these are also the most enduring and the most striking. They were created by the remarkable Zeno of Elea in about 450 BC. Zeno was a disciple of a local philosopher called Parmenides who held that the Universe was just one thing and that it consisted of just a single thing. This one thing was timeless and changeless. This led to the immediate conclusion that nothing really moves because this would require more than one thing, or state, to exist: one before the movement occurred and another after it had finished. All the movement we see, claimed Parmenides, was just an illusion on the surface of things. Deep down, the Universe was a single changeless reality.

At first, Parmenides does not seem to have been taken too seriously. This is not a convincing view to defend in the face of motion here, there, and everywhere seemingly witnessing to the contrary. Indeed, his views about motion seemed to be just simply wrong and rather obviously so.

This is where Zeno came to his teacher's rescue. Influenced no doubt by what he had learned from Parmenides, and appreciating the subtlety of his views, he set out to show that the idea that motion is possible is not quite so obvious at all. He produced four arguments to show that motion is impossible. These arguments, or 'Zeno's Paradoxes' as they became known, were never refuted in ancient times and continue to attract serious attention even today. The first two[21] draw on the mystery of the infinite in order to begin with something so simple that no one would think it remotely controversial and draw from it a conclusion that no one would believe.

Zeno's First Paradox purports to show that motion is impossible because if you want to walk from one point to another you must first cross half the distance, then half the remaining distance, then half the remainder, and so on (see Figure 1.7). If the two points are one kilometre apart you will first reach ½ km from your start, then ¾ km from your start, then $\frac{7}{8}$. After you have taken N steps of the journey you will have gone a distance equal to $1-\frac{1}{2^N}$ kilometres. No matter how big N becomes, Zeno argues, this distance will *always* be less than 1 and you will never arrive at your destination! The same argument applies to any motion over any distance, however small. You could only arrive if you took an infinite number of steps. Zeno rejected the infinite and so he rejected motion as well.

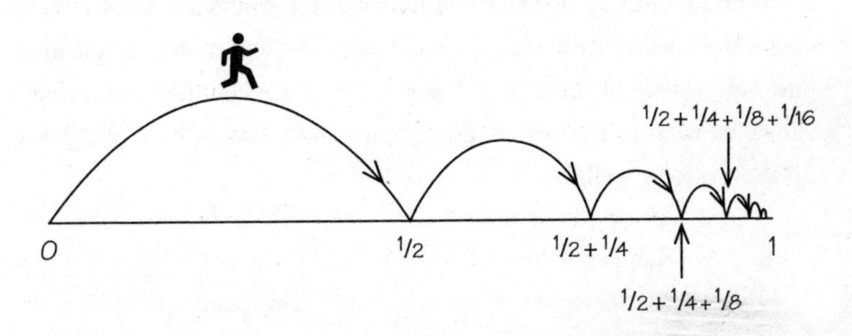

Fig 1.7 *Zeno's First Paradox: each step taken is half the length of the previous one. If you aim to walk a distance of one metre then you must first traverse half a metre, then a quarter of a metre, then an eighth of a metre, then a sixteenth of a metre, and so on forever, and it will take you an infinite time to complete the walk. So you will never arrive, Zeno argues!*

Zeno's Second Paradox creates the scenario of a race between the famous athlete Achilles and a slower rival that legend has turned into a tortoise. Achilles starts at position 0 while the tortoise, who can only run at half the speed of Achilles, has a head start and begins at the same time from the one kilometre mark.

You might have thought that Achilles, running twice as fast as the tortoise, would overtake it at the 2 kilometre mark. However, when Achilles reaches the 1 km mark the tortoise has already gone $1+\frac{1}{2}$ km; when Achilles reaches the $1\frac{1}{2}$ km point the tortoise has reached $1+\frac{1}{2}+\frac{1}{4}$ km; and so on. When, after N steps, Achilles reaches a distance $2 - \frac{1}{2^{N-1}}$ from the start the tortoise is still in the lead because it is at a distance $2 - \frac{1}{2^{N+1}}$ from the start. No matter how big N (the number of divisions of the journey) becomes, Achilles *never* overtakes the tortoise!

From these examples it would be easy to think that the infinite means nothing but paradox and a world that is restricted to the finite makes life simpler for everyone. Alas, finiteness has its own paradoxes that did not escape the attention of the ancients. A universe that is finite appears to need an edge, so what will happen if a stone is thrown across the boundary of the finite universe? Does it cease to exist as it passes the border? And what is on the outside? And if time is not infinite how can it begin and end? As we shall see, modern cosmology allows us to cure the ills of the infinite and assuage these paradoxes of the finite as well.

chapter two

Infinity, Almost and Actual, Fictitious and Factual

'We know that the infinite exists without knowing its nature, just as we know that it is untrue that numbers are finite. Thus it is true that there is an infinite number, but we do not know what it is.'

Blaise Pascal[1]

DARKNESS AT NOON

'Mathematics is the science of the infinite'

Hermann Weyl[2]

All of the intimations of the infinite that we have introduced provide attractive reasons why humans are drawn to the idea of things without end. Our minds evolved to make fruitful use of the ability to recognise pattern so as to be able to imagine the future rather than merely to experience it. The unfettered extension of that ability results in our imaginings of sequences without end. Nor do we have to run our imaginations through all the steps. We can jump in at the deep end and think about what lies beyond.

Equally important is our innate sense of the transcendental: that there is something greater than ourselves – perhaps something greater than everything. Here, perhaps the runaway imagination gets a little help from the cosmic environment in which we find ourselves.

If we look out into the dark night sky from a campsite far from the city lights, we recapture something of the ancient experience of being a human being on planet Earth. The dark night sky, threaded with beads of light from the distant stars and planets, must have been an impressive source of wonder. Its regularities enabled time to be kept and days and months to be clocked. But why is the sky so dark? At first you might think that the answer is simply that the Sun goes down. But this is not sufficient to explain what is seen. Looking out into the Universe should result in every line of sight ending on the surface of a star, just as a view into dense forest ends always on the trunk of a tree. The result should be a night sky that looks everywhere like the shining surface of a star. Why is it not so? The answer is provided by modern astronomy. The sky is dark because the Universe is so old and so big and hence so empty. It must be billions of years old in order for stars to form and provide the life-supporting elements that are needed for atom-based life to evolve and persist. And because it is expanding the Universe necessarily grows billions of light years in size and has a very low density of matter – barely one atom in every cubic metre of space on the average. If all this matter was turned into radiation we would hardly notice. It is far too small to illuminate the night sky.

We see that there is an unusual chain of connections. The huge age and size of the Universe are needed for life to be possible within it. If conscious life does arise on a planetary surface, then it must find itself in a cosmic environment that inspires thoughts about the largeness and transcendental nature of space beyond. Does the vastness of dark and almost empty space go on forever or not?

Early thinkers grappled with this question and came up with many possible answers. These were not really based on any type of observation of the Universe – how could they be in an age without telescopes

– but on trying to produce a harmonious and self-consistent story that placed what could be seen around us in a meaningful context. Everything needed to have a place and a significance. Chance and randomness were not valid forms of explanation. They were shorthand for the actions of the gods.[3] However, opinions differed as to what was meaningful. The most influential of the early Western thinkers was Aristotle who believed that the Universe of matter was finite yet surrounded by an infinite void. In many ways this was just another opinion among many, albeit with a particular philosophical reason behind it. The Earth needed to be at the centre of things and only a finite system allowed the existence of a single centre. Yet, Aristotle had other influential things to say about the whole concept of infinity. As is often the case with Aristotle, they make precise the perceptions of things that would be the commonsense view of the world.

One of the earliest and most influential considerations about infinities was that of the difference between what Aristotle called 'actual' and 'potential' infinities. In general, the Greeks had a fear of admitting infinity into their system of mathematics in the same way that they avoided the introduction of the zero.[4] Zero threatened to allow 'nothing' to be regarded as something and so smuggle a contradiction like a Trojan horse into their system of logic. Infinity looked rather similar. It could not be combined with other numbers in the usual way – add one to infinity and it is still infinity – and its appearance seemed to be disturbingly linked to the concept of 'nothing'. Divide any number by nothing and the answer is infinity. This is the route by which the early Indian mathematician, Brahmagupta, having defined a zero symbol in his decimal system of arithmetic, was able to write down equations like

$$\text{Infinity} = 1/0 \text{ and } 0 = 1/\text{infinity}$$

in AD 628, unperturbed by any wider philosophical ramifications of the concepts of either zero or infinity.

The potential to create infinities by continually increasing the size of something, or by diminishing it, was a very ancient intuition. Amongst the pre-Socratic philosophers, it is elegantly stated by Anaxagoras (500–428 BC), who tells us that

> 'There is no smallest among the small and no largest among the large but always something still smaller and something still larger.'

The Greeks came very close to accepting the infinite into their mathematical system, but they could never quite take the final leap needed to embrace it. They were happy to consider processes that continued without stopping, but were influenced by the famous paradoxes of Zeno who signalled the dangers inherent in following the path all the way to infinity. As a result, the appearance of the infinite was saddled with the expectation of paradox and puzzlement. It would be a brave philosopher who dealt with the problems of infinity head on.

A PURELY ARISTOTELIAN RELATIONSHIP

'We hold these truths to be self evident'

The American Declaration of Independence

Aristotle was never one to avoid any area of human inquiry because it was problematic. As befits the student of Plato and the supervisor of Alexander the Great, he had something to say about every subject under the Sun, and many beyond it too. His view about infinities was succinctly and plainly put:

> 'The infinite has a potential existence . . . There will not be an actual infinite'[5]

What did he mean? We are familiar with sequences that can be continued without end. Count the whole numbers, 1, 2, 3, 4, 5, 6, 7, . . . There is no largest number (if you think there is, just add one to it) and we see an example of a potential infinity: a sequence without end. Aristotle recognised this as a form of process, an infinity that was never actually completed by any counter, a listing that was always actually finite despite its asymptotic promise of infinity to come. It could never be grasped and viewed as a whole. Many other examples can be found of such limitless sequences. They don't all have to disappear off into the future, they can arrive from the past, with beginning or cause. Think of all the negative numbers – an infinite sequence that ends with –1 – that runs . . . –6, –5, –4, –3, –2, –1. Again the infinity to the left is potential in nature, we cannot see the whole sequence or dredge up its beginning, for it has none.

Aristotle argued that there could not be any object that was infinitely large. Yet, he did not want to outlaw infinite time sequences because otherwise time would have to have a beginning and an end, and the sequences of negative and positive numbers would have to have a first and a last member. It would also mean that there was a smallest piece of matter that could not be divided in half and that seemed as odd as the idea that there might be a biggest number. He explains in more detail:

> 'A thing may be said to be either potentially or actually, and a thing may be infinite either by addition or by division. Now we have said that no magnitude is actually infinite, but magnitudes are infinite by division (for the thesis that there are indivisible lines is easily refuted), and therefore they must be potentially infinite. But their being potentially infinite must not be understood as implying that they will at some time be actually infinite in the same way as what is potentially a statue

> will at some time be a statue. For being has many senses and the sense in which a thing is infinite is the sense in which there is a day or a contest, namely by one thing coming into existence after another. For indeed in these cases too we may distinguish between potentiality and actuality: the Olympic games exist both in the sense that they are able to take place and in the sense that they are taking place.'[6]

There is something a little odd about Aristotle's distinction between potential and actual infinities here. When we talk about the distinction between, say, someone being a potential Prime Minister and an actual Prime Minister we imply that it is physically and logically possible that this someone should actually become Prime Minister. Aristotle gives a similar example of the Olympic games. It can potentially occur in Athens means that it could actually happen in Athens. Yet, Aristotle seems to be making infinity an exception to this way of thinking – perhaps the only exception – because his potential infinities can never become actual. He would look at a piece of wood and admit that it could be cut into an unlimited number of pieces, but the process of cutting into an infinite number of pieces could never be completed. He means of course that it could never be completed in a finite period of time. In a later chapter we shall see that modern physics introduces some unusual possibilities in this respect, and different observers can see the same process taking very different periods of time. Aristotle maintains that it is impossible to complete an infinite series of distinct tasks.

This view leads Aristotle to take a very different view to the earlier tradition that viewed the infinite as a nebulous 'everything' in which everything conceivable was contained. This, he thinks, is a confusion and in reality the infinite is exactly the opposite:

> 'The infinite turns out to be the opposite of what they say. The infinite is not that of which *nothing* is outside, but that

> of which there is *always something* outside. That of which nothing is outside is complete and whole. By contrast, that of which something, whatever it might be, is absent is not everlasting.'[7]

Thus he thinks that the infinite is imperfect and incomplete.[8] For Aristotle the infinite is not, as we might think, some all-encompassing transcendent notion that contains everything. Quite the contrary: pieces of matter can be divided without limit and so the infinite is immanent within every piece of matter in the world. Thus the infinite is part of the ultimate nature of matter, irrespective of the shape or form that it takes in particular objects. Hence this ultimate nature is beyond our ability to understand and the infinite is a necessary part of that unbreachable unknown into which the human mind cannot venture and from which it cannot return.

In order to understand more fully the direction of his thinking one has to recall something broader in Aristotle's conception of the world. Everything had to have a goal, or an end, in order to give it meaning and significance. This goal, or 'final cause' as he called it, acted like a magnet in the future. So, potentially infinite sequences were unsatisfactory because they lacked that final step, or goal, that would bring about their completion and provide their true meaning and significance. We could not know the explanation of anything that was infinite. It is because the world is finite and contains only things that are also finite that the whole enterprise of human inquiry into the nature of the world is possible. As the philosopher Jonathan Lear puts it:

> 'But we *can* know the causes of a thing; therefore they must be finite. If those properties which make a substance what it is were infinite in number, then the substance would be unknowable. But we *can* know what a substance is, therefore there are only finitely many properties in its definition.'[9]

So much for Aristotle's attitude to space and matter, what did he believe about time? Here he thinks differently. He does not exhibit the fact that each interval of time is a little encapsulation of the infinite because it could be subdivided into smaller and smaller parts without limit. Instead, it is simply the unending character of the flow of time that makes it potentially infinite in his sense. The infinite future is never reached or encompassed within the finite world. Aristotle believed that the world and the celestial circular motions were uncreated and eternal, although the amount of matter and the extent of the space in which it dwells are finite. For Aristotle, time was a measure of change, of things happening. It was not like Plato's fixed arena in which events may or may not be played out. Time could not exist if there was nothing happening or no mind to measure its passing. What Aristotle means here is not that observers and living beings somehow create or bring time into being. Rather, that without their presence to measure or experience change there would not be a complete conception of what time is.[10] Thus, if the world has always existed, there can be no measurement of its infinite age. However, all this is not entirely satisfactory. Aristotle seems perfectly happy to assume that the world has an infinite age. His defence of the idea is that if we assume that there was a first moment then we can always, say, halve that moment and we have an earlier moment.

To modern readers Aristotle's ideas seem curiously divorced from the realms of science and observation which we have learnt to look to for guidance about the likely age of the world. But in Aristotle's time there were no telescopes, no microscopes, no understanding of the size and age of the solar system. He sought to devise a philosophical system that was coherent and in which everything seen had meaning. Infinity was a challenging idea and Zeno had displayed all too clearly that to handle it wrongly could undermine our entire understanding of time, and change. Aristotle bravely erected the first far-reaching attempt to grapple with these problems. The distinction he made between actual and potential infinities was clear and simple. It was one that would last for thousands of years and convince some of our greatest mathematicians of its cogency.

INFINITY AND GOD

'When we've been there ten thousand years,
Bright shining as the Sun,
We've no less days to sing God's praise
Than when we'd first begun.'

Harriet Beecher Stowe's addition to
John Newton, *Amazing Grace*[11]

One might have thought that a constant tension in the quest to grasp the infinite by human thought would have been the suspicion that the place of God was being usurped by anyone claiming to comprehend the infinite. Yet, this seems to have been less of a problem than the concern that the infinite might be something that even God could not comprehend. The simplest example of the never-ending list of natural numbers (1, 2, 3, 4, . . .) was just such a challenge. So seriously does this idea seem to have been taken that St Augustine, the most authoritative of the early Church fathers, devotes considerable space in his writings to answering it. In a section headed '*The answer to the allegation that even God's knowledge cannot embrace an infinity of things*', he argues that,

> 'there is the assertion that even God's foreknowledge cannot embrace things which are infinite. If men say this, it only remains for them to plunge into the depths of blasphemy by daring to allege that God does not know all numbers. It is certainly true that numbers are infinite . . . Does this mean that God does not know all numbers,

> because of their infinity? Does God's knowledge extend as far as a certain sum and end there? No one could be insane enough to say that.
>
> . . . And the prophet says of God, "He produces the world according to number"; and the Saviour says in the Gospel, "Your hairs are all numbered."
>
> Never let us doubt, then, that every number is known to him "whose understanding cannot be numbered." . . . And so, if what is comprehended in knowledge is bounded within the embrace of that knowledge, and thus is finite, it must follow that every infinity is, in a way we cannot express, made finite to God, because it cannot be beyond the embrace of his knowledge.'[12]

Augustine's idea of the humanly infinite appearing finite to the mind of God is an interesting speculation on his part. In modern mathematics just such a trick is frequently applied in order to make an infinite region finite and more easily visualised and representable in a diagram. A mathematical transformation – the numerical equivalent of translating from English into another language – can be applied which brings infinity in to a finite point.[13] The is not a magic recipe for removing all infinities from consideration because, by the same token, there are transformations that take finite points into infinite ones.

This threat to God from the infinite would later re-emerge with a more subtle complexion. In the Renaissance, God's attributes were primarily those of Being and Infinity. If the infinite were to be tamed by mathematicians or philosophers so that it could be encompassed by a mere sheet of paper, then what would be left for God to be? In such circumstances it is important that theologians do not identify God with the infinite in an essential way. That is not to say that they should argue that God is finite – that would only make the problem worse – but only to deny that just because we can understand an attribute of God in some way, that God is diminished.

Not all philosophers saw the infinite as a theological problem. Nicholas of Cusa embraced the infinite as a place where opposites could be reconciled. He saw its paradoxical properties as an object lesson in the type of explanation that was needed to understand truth and reality, of which our own conceptions are merely approximations. He liked to use mathematics as an analogy of the problems of theology. Thus, for example, when we are dealing with finite lines there is a clear distinction between a straight line and a curved line, as we can see by looking at the line and the circles in Figure 2.1.

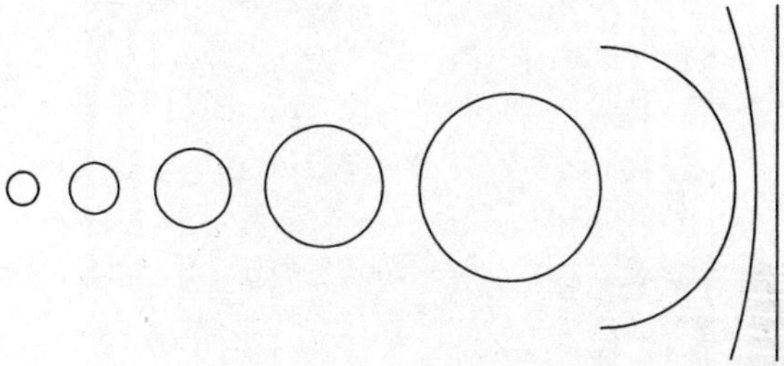

Fig 2.1 *A line and a circle. As a circle increases in diameter it looks more and more like a straight line locally.*

Yet, if the circle were increased in size so as to have an infinite diameter, then its surface would be indistinguishable from an infinite straight line.

The seventeenth-century French philosopher, mathematician, and scientist Blaise Pascal (Figure 2.2) introduced his characteristically original slant on the theology of the infinite by using the fact that if any finite quantity, however small, is multiplied by infinity then the result is infinite.

Pascal, although an ardent theist, wanted to consider a pragmatic argument that might convince an atheist that belief in God was the best option to bet on. Pascal was one of the pioneers of modern probability theory and liked to consider the role played by uncertainty and chance in practical problems. 'Either God exists or not,' he pronounces, and 'reason cannot decide for us one way or the other;

we are separated by an infinite gulf'. So what should we do? We can believe in God by faith, as Pascal chose to do, but there is another approach he argues. You are laying a bet. How should you play? Pascal asserts that you have two options: believe in God's existence or disbelieve.[14] And there are two possible situations that may exist: God exists and God does not exist. What are the results of your possible choices?

Fig 2.2 *Blaise Pascal (1623–62).*[15]

Pascal argues that the best bet is to believe in God because the infinite gain that will result if he does exist will always outweigh the purely finite loss of time that will result if he does not (Figure 2.3). Likewise, the atheistic position is the worst bet because if God does exist the loss will be infinite, whereas if he doesn't the human gain will be only finite.

Pascal confronted the problem of what became known as the 'double infinity'. He maintained that infinity was encountered everywhere in the world, but it was not possible for the human mind to apprehend it fully. One of the two infinities was the potential infinity of being able to increase quantities without limit, whether they be

	God exists	God doesn't exist
believe in God	infinite gain	finite loss
disbelieve in God	infinite loss	finite gain

Fig 2.3 *The pay-offs that result from two different beliefs about God's existence in the two different situations that might actually be the truth.*

simply the natural numbers we count with or physical quantities, like speed of motion, which he (wrongly[16] in retrospect) believed could increase without bound. The other was that of infinite smallness that was present in all things because of the potential for indefinite subdivision and, in the case of motion, for a speed to be halved in value in a never-ending succession of steps:

> 'Thus there are properties common to all things, and the knowledge of them opens the mind to the greatest wonders of nature. The principal one includes the two infinities which are to be found in all things, infinite largeness and infinite smallness.'[17]

He goes on to apply this consideration to the nature of space and time, arguing that

> 'No matter how large a space is, we can imagine a larger one, and still a larger one than this, and so on infinitely, without ever arriving at one which could no longer be increased. And conversely, no matter how small a space may be, we can still think of a smaller one, and so on infinitely, without ever reaching one which is indivisible because it no longer has an extent.
>
> The same applies to time.'[18]

Thus he concludes that the fact that the double infinity of the arbitrarily large and the arbitrarily small inhabits space, time and motion means that it is potentially present in all things. Indeed it is the hallmark of Nature, he suggests, that

> 'since nature has engraved her own image on all things, and that of her author on all things, they almost all share her double infinity.'[19]

From our modern perspective these arguments are less than convincing. The infinity of space is assumed in order to assert its presence. If the Universe were finite in volume, then this argument for the ever-increasing nature of space would fail. Likewise the argument from motion: we now know that Nature is fashioned so that there is a finite maximum speed of all motion that transfers information. It is the speed of light in a perfect vacuum. The third argument, for the infinite divisibility of space and time, is now also dubious. It is likely that there is a minimum time interval and length defined by the constants of Nature.[20]

For Pascal, these two infinite extremes were beyond human conception. They were potential infinities in Aristotle's ancient sense. We were in between the two infinities, but both were equally inaccessible to our minds. One, the infinitely large, seems naturally beyond our reach, but we should not be tempted to think that the infinitesimally small is any closer to our understanding just because we happen to be greater than it:

> 'Let us realise our limitations. We are something and we are not everything. Such being as we have conceals from us the knowledge of first principles, which arise from nothingness, and the smallness of our being hides infinity from our sight.'[21]

In this course Pascal follows Galileo who had written in a very similar way that we must

> 'remember that we are among infinities and indivisibles, the former incomprehensible to our understanding by reason of their largeness, and the latter by their smallness. Yet we see that human reason wants to abstain from giddying itself about them.'[22]

Pascal's arguments about the infinity of the natural numbers remain, though. He uses this undeniable infinity as an argument for the fact that we can know that an infinity exists without knowing its nature. We can contemplate it, but not fully conceive of it. A subtle apologetic is being made here: an argument that we can reliably deduce things about the existence of an infinite God on the basis of finite experience, just as we can infer truths about mathematical infinity from numbers that are individually finite.

The same focus on the mystery of the potential double infinity permeating the world is made by Descartes, but he introduces a new delineation. He wants to apply a little spin-doctoring and rename the situations where there is the potential for what we see to increase without limit as 'indefinite'. By avoiding the use of the term 'infinite' to describe the potential infinities latent in 'the extension of the world, the division of the parts of matter, the number of the stars, and so on', he believes that 'we will never be involved in tiresome arguments about the infinite'.[23] Having parcelled up all these topics about potential infinities that are discussable on the basis of approach from observable finite quantities, Descartes wants to dispatch all other discussion about actual infinities to the dustbin of unwarranted speculation. Our human limitations make it impossible to discuss the subtleties of the infinite. Thus, he takes up the very examples that would later become crucial in a rigorous formulation of what actual infinities are in mathematics, and argues that we must not concern ourselves with them,

> 'For since we are finite, it would be absurd for us to determine anything concerning the infinite; for this would be to attempt to limit it and grasp it. So we shall not bother to reply to those who ask if half an infinite line would itself be infinite, or whether an infinite number is odd or even, and so on. It seems that nobody has any business to think about such matters unless he regards his own mind as infinite.'[24]

And the punch-line, which explains why we are discouraged from pursuing the study of the infinite, is:

> 'so as to reserve the term "infinite" for God alone. For in the case of God alone, not only do we fail to recognise any limits in any respect, but our understanding positively tells us that there are none.'[25]

In the face of this challenge it was important that subsequent philosophers, at least in France, drew a distinction between metaphysical infinity and mathematical infinity so as to be able to study geometry and arithmetical series unrestrained by these Cartesian admonitions. And indeed they did.

Western thinkers after the Greeks generally accepted that God was infinite in many, if not all, respects. In the Christian tradition the main difference in outlook that then followed was that of the pantheists, like Spinoza and Hegel, who held that there was no distinction between God and the physical Universe: the Universe was the totality of everything that exists. Thus Spinoza would then affirm that this totality must be infinite and hence God must be infinite or he would be limited by something about the Universe that is infinite. The other historical position, which many modern theologians would adopt still, is that of panentheism, which maintains that God wholly transcends the physical Universe, whether it be finite or infinite. That is, the physical Universe is a proper subset of God. Theism, by contrast, goes a step further and regards God as wholly other than the physical Universe.[26] Obviously, if the physical Universe is finite – as is entirely possible in modern cosmology – then the infinite nature of God reinforces this otherness in an important way but it is not the essence of it. The otherness of God would be assumed even if the Universe were infinite.[27]

A LITTLE KANT

'For the intellect is to truth as an inscribed polygon is to the inscribed circle. The more angles the inscribed polygon has the more similar it is to the circle. However, even if the number of angles is increased ad infinitum, the polygon never becomes equal to the circle unless it is resolved into an identity with the circle.'

Nicholas of Cusa[28]

Immanuel Kant's enduring contribution to philosophy was to argue for a clear distinction to be made between what we might call 'true reality' and 'perceived reality'. The act of using our minds and sense to apprehend the nature of true reality produces a change in its nature (and even if it doesn't, we can never know that it doesn't[29]). We like to think of ourselves as if we are bird-watchers in a perfect hide, who can observe and study the world without disturbing it. Kant (1724–1804) argued that this is impossible in principle. Our minds possess certain pigeon-holes for understanding things and the information we gather about the world inevitably finds its way into these slots. This introduces a brake on all claims to know anything about the ultimate nature of things or to answer great philosophical questions about the existence of God and the meaning of life. The role of our categories of thought in processing the raw character of the Universe into human knowledge may, of course, be a harmless detail, but maybe not.[30]

Kant believed that the Universe was infinite, both in extent and in the multiplicity of possibilities that it can give rise to. In these respects it was a reflection of God. However, beliefs about such things are not the same as knowledge. Kant placed these infinities in the realm of the true realities, but all our perceptions and understanding of them were necessarily finite, constrained by our human modes of perception. Thus for Kant actual infinities do exist, but they can only be understood as finite perceptions, or 'phenomena' as he called them.

In some respects Kant is comparable with Nicholas of Cusa (1401–64) who also separated perceived reality from actual reality and believed that the Universe was both infinite in spatial extent and in variety. In these attributes it reflected the inexhaustible character of God, but this made God inconceivable and unrepresentable in finite form, an idea that is also familiar in Islamic and Jewish thought.

Kant's dominant place in European philosophy, especially in German schools, meant that his influence was felt far beyond the world of professional philosophers. Nineteenth-century scientists and mathematicians were strongly influenced by his views, and in 1831 great mathematicians like Gauss spurned the consideration of actual infinities just as robustly as the ancients. But a revolution was coming.

chapter three

Welcome to the Hotel Infinity

'I am returning this otherwise good typing paper to you because someone has printed gibberish all over it and put your name at the top.'

Professor of English, Ohio University

HOTELS

'Three guys go into a hotel, each with $10 in his pocket. They book one room at $30 a night. A short while later a fax from headquarters directs the hotel to charge $25 a night. So the receptionist gives the bellhop $5 to take to the three guys sharing a room. Since the bellhop never got a tip from them and because he can't split $5 three ways, he decides to pocket $2 and give them each one dollar back. So each of the three guys has now spent $9 and the bellhop has $2, for a total of $29. Where's the extra dollar?'

Frank Morgan[1]

Hotels are memorable places. Ironically, the worse they are the more memorable they are. The BBC's classic situation comedy *Fawlty Towers* began after John Cleese spent some time in the seaside town of Torquay at a strange hotel that was run by an even stranger man. Of course, in the TV drama, the most intriguing character of all, the Major, lived there all the time.

Nor is Torquay unique, a few years ago the London *Times* carried a story from a businessman who had checked into his room early at a New York hotel to find that the previous occupant had died in his sleep and was still in the bed. He rushed to the front desk to tell the concierge breathlessly that 'there's a dead body in room 123'. Without looking up, the concierge reached behind him to the key board and calmly said, 'Take 124 instead.' The worst hotel I ever stayed in had a foot-shaped hole through the door and a large shower room with rotting wooden door that contained the toilet and all the accompanying electrical fittings. All were saturated on use of the shower and it took about an hour for the water to drain away and only a little longer for me to run away as well.

The thing about hotels is that everyone is a stranger and there can be an unnerving lack of knowledge about the number and nature of the other guests. Anonymity is preserved and life becomes totally numerical: there are room numbers, floor numbers, phone codes, times for breakfast and check-out, taxi pick-up times, internet log-on numbers, credit card numbers at check-in as well as check-out, the number of bottles of water taken from the minibar, exchange rates, and a bill of astronomical magnitude when you depart – you can count on it. To top it all, there are mirrors everywhere to create never-ending mutual reflections. What better place to unleash the infinite?

EXPERIENCES OF THE HOTEL INFINITY

> 'Standing among savage scenery, the hotel offers stupendous revelations. There is a French widow in every bedroom, affording delightful prospects.'
>
> Gerard Hoffnung[2]

Fig 3.1 *Picture from the opening scene set in the Hotel Infinity from the author's play* Infinities, *directed by Luca Ronconi, performed by the Teatro Piccolo in Milan in 2002 and 2003.*[3]

The essence of the infinite is beautifully captured by the story of the Infinite Hotel that is attributed to the great German mathematician David Hilbert.[4] Hilbert was eccentric in his own rather severe way. An oral tradition of Hilbert anecdotes grew up in his lifetime in a similar manner to those surrounding the peculiar British physicist Paul Dirac. One story tells of how one of Hilbert's students committed suicide, after failing to solve a challenging mathematical problem. Hilbert was asked by the student's family to talk at the funeral. In his address at the graveside he explained that the maths problem that had caused the young man's death was in fact fairly simple. The student, he went on, had simply looked at it in the wrong way.[5]

Not surprisingly then, any hotel Hilbert had in mind is likely to be a little odd. In a conventional hotel there are a *finite* number of single rooms (see Figure 3.1). If they are all taken, then there is no way you can be accommodated at the hotel without evicting one of the existing guests from their room. When it's full it's full.

At the Hotel Infinity things are different. Suppose that you turn up at the check-in counter of the Hotel Infinity only to find that the infinite number of rooms (numbered 1, 2, 3, 4, . . . and so on, forever) are all occupied. The receptionist is perplexed – the Hotel is full – but the manager is unperturbed. No problem, he says: move the guest in room 1 to room 2, the guest in room 2 to room 3, and so on, forever. This leaves room 1 vacant for you to take and everyone still has a room!

You are so pleased with this service that you return to the Hotel Infinity on the next occasion that you are in town, this time with an *infinite* number of friends for the ultimate reunion. Again, this popular hotel is full, but again, the manager is unperturbed. We can easily accommodate an unexpected party of infinity, he explains to the nervous receptionist. And so he does, by moving the guest in room 1 to room 2, the guest in room 2 to room 4, the guest in room 3 to room 6, and so on, forever. This leaves all the odd-numbered rooms empty. There are an infinite number of them and they are free to accommodate you and your infinitely numerous companions without leaving anyone out in the cold. Needless to say room service is a little slow at times to some of the high-numbered rooms.

The day after the infinite contingent of unexpected guests have been accommodated, the disgruntled guests in the even-numbered rooms all decide to leave. They are fed up with being constantly moved around by the crazy manager and spending all their time queuing for everything.

The manager is very upset that half of the hotel's rooms (all the even numbers) are now empty. He has to supply statistics on the occupancy of the Hotel and 50 per cent occupancy is a failure. Unless things look up he is facing closure. As a demanding traveller you have begun to get the idea how things work at this hotel now. You don't want to see such a flexible establishment close down, so on hearing of the manager's problem you suggest that the remaining guests just be moved closer together to get rid of the unoccupied rooms. You propose that they leave the guest in room 1 alone, move the guest in room 3 to room 2, 5 into 3, 7 into 4 etc. In the end all the rooms are filled again even though no new guests have arrived. The manager is delighted.

The next day the manager is depressed again. His hotel is a member of a chain that has an infinite number of sister hotels, one in each galaxy in the infinite Universe. However, the intergalactic business has not been doing well and huge closures ('restructuring') are becoming necessary.

He explains that this has brought good news and bad news. The *good* news is that the bosses have been so impressed with some of the manager's recent efforts to accommodate last-minute business that they have decided that their best chance of financial survival is to sack all the other managers (cutting an infinite amount off the salary bill) and close every hotel in their infinite chain except his. The *bad* news is that the infinite number of existing guests staying in each one of the infinite number of other members of the Hotel Infinity chain are to be moved to his hotel. He is faced with suddenly having to find rooms for an influx of guests from infinitely many other hotels, each of which has infinitely many guests, when his own hotel is already full!

The resourceful manager had begun by finding room for one extra guest in a full hotel, then room for an infinite number of extra guests in a full hotel, and now he is being asked to find room for an infinite number of travel parties, each of which contains an infinite number of guests. What can he do? They will start arriving soon.

Everyone in the hotel is set to work on the problem. Lots of crazy suggestions are made, all unsuccessful, but then someone comes up with a promising proposal.[6]

Why not try this strategy? Leave the guest in room 1 alone, move the guest in room 2 to room 1001, room 3 to 2001, room 4 to 3001, and so on. Now put the newly arriving guests from Hotel 2 into rooms 1002, 2002, 3002, . . . etc of our Hotel Infinity, the guests from Hotel 3 into rooms 1003, 2003, 3003, etc. At first this seems to be the answer. It looks as if everyone will have a room and there is a definite system for allocating them. Then the receptionist notices something that nearly gives the manager a heart attack. What is going to happen to the guests from the 1001st hotel? They have nowhere to go because the guests arriving from the first 1000 hotels in the chain would already have occupied all the rooms. They are back to square one[7] and already there are lots of spaceships on the horizon.

Next someone suggests putting guests from Hotel 1 in rooms 2, 4, 8, 16, and so on, multiplying by 2 each time; the guests from Hotel 2 in rooms 3, 9, 27, 81, and so on, multiplying by 3 each time; and so on forever. But the manager realises there are problem rooms which would end up with more than one guest – for example room 16 would receive the fourth from Hotel 1 as well as the second from Hotel 3. You need to make sure the assignment scheme sends only one person into each room. One needs a way of assigning the two numbers which is unique.

Then a student kitchen worker, who has just finished his first year of a maths course, suggests using prime numbers[8] (2, 3, 5, 7, 11, 13, 17, 19, 23, . . . there are an infinite number of them) because any whole number can be expressed as a product of prime factors in only one way. For example $8 = 2 \times 2 \times 2$ and $21 = 3 \times 7$ and $35 = 5 \times 7$. The manager is intrigued. Some of his old maths lessons are coming back to him. He listens patiently, thinks carefully, and then makes an announcement to all the staff. Here is the plan: put the infinite number of guests arriving from Hotel 1 into rooms 2, 4, 8, 16, 32, . . . ; those arriving from Hotel 2 into rooms 3, 9, 27, 81, . . . ; those arriving from Hotel 3 into rooms 5, 25, 125, 625, . . . ; those from Hotel 4 into 7, 49, 343, . . . and so on. No room can ever be assigned more than one guest because if p and q are different prime numbers and m and n are whole numbers then p^m can never be equal to q^n.

Trying things out, the staff notice that things can be simplified a little and the recipe for room allocations applied very easily with the aid of their desk calculator. Just put the guest arriving from the mth room of the nth hotel into room number $2^m \times 3^n$; for example the 6th guest from the 4th hotel goes into room $2^6 \times 3^4 = 64 \times 81 = 5184$. No room could have two occupants.

But still the manager is not happy. If this plan is implemented there will be a huge number of unoccupied rooms! In the student's original plan all the rooms with numbers like 6, 10, 12, whose numbers are not powers of prime numbers, will be left empty. In the manager's plan all rooms which cannot be written as number $2^m \times 3^n$ will be left empty. Reduced to desperate measures, the manager calls one of his old school friends who now runs a management consultancy which offers advice on how to run businesses efficiently. For an infinite fee his agency quickly comes up with a new suggestion that is much more efficient:

Draw up a table with bracketed pairs of numbers denoting the old room number of the arriving guest and the old hotel number from which they are coming. So, for example, the entry on the 5th row of the 4th column labels the guest from the 5th room of the 4th hotel (Figure 3.2).

(1,1)	(1,2)	(1,3)	(1,4)	. . .	(1,n)
(2,1)	(2,2)	(2,3)	(2,4)	. . .	(2,n)
(3,1)	(3,2)	(3,3)	(3,4)	. . .	(3,n)
(4,1)	(4,2)	(4,3)	(4,4)	. . .	(4,n)
(5,1)	(5,2)	(5,3)	(5,4)	. . .	(5,n)
. . .	. . .	. . .	. . .	. . .	. . .
(m,1)	(m,2)	(m,3)	(m,4)	. . .	(m,n)

Now there is a simple rule to deal with the new arrivals using the agency's table. When the guests arrive, tell your check-in staff to put the guest from (1,1) in room 1; the guest from (1,2) in room 2; the guest from (2,2) in room 3; the guest from (2,1) in room 4. This deals with all the guest from the top left-hand 2 × 2 square in the table.

(1,1) to room 1	(1,2) to room 2	(1,3)	(1,4)	. . .	(1,n)
(2,1) to room 4	(2,2) to room 3	(2,3)	(2,4)	. . .	(2,n)
(3,1)	(3,2)	(3,3)	(3,4)	. . .	(3,n)
(4,1)	(4,2)	(4,3)	(4,4)	. . .	(4,n)
(5,1)	(5,2)	(5,3)	(5,4)	. . .	(5,n)
. . .	. . .	. . .	. . .	. . .	. . .
(m,1)	(m,2)	(m,3)	(m,4)	. . .	(m,n)

Now do the 3 × 3 square. Put the guest from (1,3) in room 5, the guest from (2,3) in room 6, from (3,3) in 7, (3,2) in 8, (3,1) in 9. The 3 × 3 square in the top left is now done.

(1,1) to room 1	(1,2) to room 2	(1,3) to room 5	(1,4) . . . (1,n)
(2,1) to room 4	(2,2) to room 3	(2,3) to room 6	(2,4) . . . (2,n)
(3,1) to room 9	(3,2) to room 8	(3,3) to room 7	(3,4) . . . (3,n)
(4,1)	(4,2)	(4,3)	(4,4) . . . (4,n)
(5,1)	(5,2)	(5,3)	(5,4) . . . (5,n)

Fig 3.2 *The algorithm emerges in the lobby of the Hotel Infinity in scene one of* Infinities.

The manager is excited, but is there going to be enough room he asks? Yes;[9] the maths student reappears to show him that not only would every arriving guest be accommodated in their own room, but not a single room would be left empty. Occupancy is back to 100 per cent!

THE HOTEL INFINITY'S ACCOUNTS

'All boarding houses are the same boarding house. A single room is that which has no parts and no magnitude.

All the other rooms being taken, a single room is said to be a double room.'

Stephen Leacock[10]

Business at the Hotel Infinity is at an all-time high. Revenue is infinite, costs are infinite, but profits are infinite – this bottom line is all that the manager's accountant needs to tell him. That is, until he gets a tax bill. The Hotel Infinity's accountant has managed to ensure that their tax rate is the lowest possible – lots of intergalactic diversification and artificial tax domiciles – but no matter what the tax rate is, when you apply it to the infinite income the result is infinite.[11] 'How can this have happened?' the manager screams. 'We are ruined. Our tax liability is infinite just like our profit.' The accountant sits him down in a comfortable chair and makes him a nice cup of tea. 'Let me explain,' he says. 'Just go ahead and pay your infinite tax bill. You will find that your profits are quite undiminished. They will still be infinite.'

All does not end well. The Hotel Infinity's long-suffering owners are gradually worn down by the complexities of managing infinitely many guests from infinitely many hotels in infinitely many galaxies. They are stuck in a recession of intergalactic proportions. It is predicted to last for billions of years. They decide to escape by making a *radical* change of business strategy – an infinitely radical change. They decide to rename the chain, rebrand their products, move into a new commercial niche. They decide to become the ultimate in fashionability – the ultimate minimalist hotel. They become the Hotel Zero. Life is simpler. Now there are no rooms, no guests, no staff, no running costs (room temperature is kept at absolute zero), no losses, no problems. There's even canned music in the bar with John Cage's work *4 minutes 33 seconds*[12] continuously playing, blank-canvas modern artworks in the lobby, and a free copy of the author's *Book of Nothing*[13] handed by way of consolation to every hopeful but disappointed guest, of whom there are many, infinitely many. And on the wall the thought for the day reads

> 'If people do not believe that mathematics is simple, it is only because they do not believe how complicated life is.'

chapter four

Infinity Is Not A Big Number

'That is infinite, this is infinite, from infinite arises infinite, when infinite is subtracted from infinite, what is left is infinite'

Sanskrit Mantra[1]

AN IMMACULATE MISCONCEPTION

'Space is almost infinite. As a matter of fact, we think it *is* infinite'

Dan Quayle[2]

There is an understandable tendency to think of infinity as simply a very big number, just a bit bigger than the biggest number you can think of, always just out of reach, like the end of the rainbow. Yet, to appreciate the subtleties of the infinite it is important to appreciate that infinity is not simply a very big number. It is qualitatively (and not just quantitatively) different from any finite number (like 124,453,567,000,000,000,000,000,000,000,001), no matter how big it is. This idea that infinity is just a very very large number is what most people are likely to think. It is tempting to think that infinity is just a count that keeps on going and so is approximately rather like the biggest number you could ever think of plus a bit more.

ALBERT OF SAXONY'S PARADOX

'And when he had taken the five loaves and the two fishes, he looked up to Heaven, and blessed, and brake the loaves, and gave them to his disciples to set before them; and the two fishes divided he among them all.
And they did all eat, and were filled.
And they took up twelve baskets full of the fragments, and of the fishes.
And they that did eat of the loaves were about five thousand men.

St Mark 6:41–44

Albert Ricmerstop was born near Helmstedt in West Saxony in 1316. He was to become one of the most influential logicians of the Middle Ages, studying in Prague and Paris before becoming first the Rector of the University of Paris and then the founding Rector of the University of Vienna in 1365. Besides producing his large body of work in logic and philosophy, he also played an important role in early political interactions between Church and State by carrying a series of diplomatic missions to the Pope on behalf of the Duke of Austria. As a result, just one year after his appointment in Vienna, he was named Bishop of Halberstadt and he remained in that office until his death in 1390. To later scholars he would be known as Albert of Saxony, or simply 'Albertucius' which means 'little Albert' in order to distinguish him from Albert Magnus ('the Great'), the famous thirteenth-century theologian.

Albert was an acute thinker who played the game of medieval theology in new ways, devising procedures for determining the truth or falsity of statements or 'sophismata'[3] used in teaching and evaluating the limits of different philosophical systems. The sophismata were sentences which were in some way hard to understand, ambiguous, or paradoxical. The name of the game was to deal with the ones that rival philosophers came up with and create telling examples of your own. A sentence like 'Nothing is something' or 'Only God is infinite' or 'This statement is false' would be simple candidates. Albert was interested in the paradoxes and problems of the infinite and discusses them in his book, *Sophismata.* In the course of his discussions he provided a wonderfully incisive paradox about infinity which was later to form the basis for the definition of an infinite collection and the foundation for a rigorous discussion of actual infinities. This wasn't Albert's intention, of course, but it shows how carefully he thought about the question, and also reveals the influence of the English philosophers of the day whose use of mathematics was enthusiastically taken up and promulgated by Albert.

Albert showed that a single infinite allows you to get something for nothing – indeed, to get as much as you want for nothing. Take an infinitely long beam of wood, with a square cross-section of size 1 unit by 1 unit (see Figure 4.1). Now saw it up into cubes of equal size. You will have an infinite number of these cubes which you can now use as building blocks. Albert argues that you can use them to fill the whole of space by assembling them in a systematic way. Surround the first block by $3^3 - 1 = 26$ blocks so as to make a bigger cube of side equal to 3 units. Now surround that cube with $5^3 - 3^3 = 98$ more blocks to create a new cube of side equal to 5 units. By continuing this process using $7^3 - 5^3$, then $9^3 - 7^3$ then $11^3 - 9^3$, and so on forever, of the original blocks, you would be able to build a single cube of ever-increasing volume. The infinitely long beam that you started with can therefore be cut up and re-assembled to fill the whole of an infinite three-dimensional space!

Albert's clever example shows that even in the fourteenth century there was a clear appreciation of the curious feature of infinity, that it can be put in direct correspondence with a part of itself. The importance of Albert's example was that it destroyed Aristotle's confident dogma that there cannot exist an infinite collection of

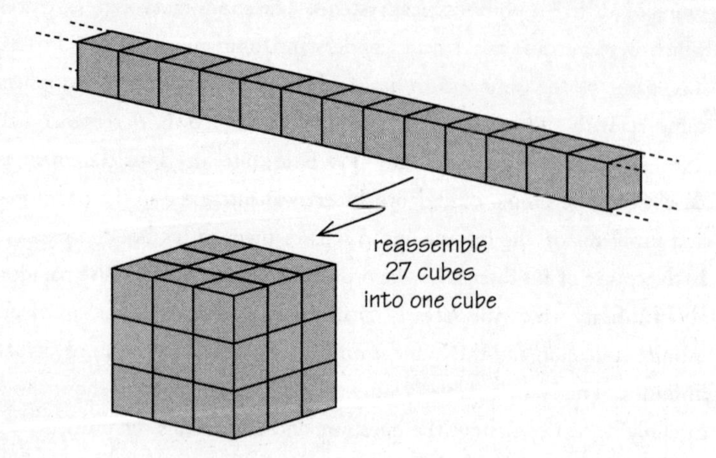

Fig 4.1 *Albert's magic process that shows how to fill the whole of infinite space by cutting up an infinitely long beam of wood that is only 1 centimetre square in cross-section and reorganising the pieces into a cube of ever-increasing size.*

things simply because it would contain a smaller subset that was also infinite and this was absurd. The example shows how such a situation can come about and there is no internal logical contradiction involved. In fact, Albert's example is cleverer than it needed to be in order to make his point, although one can imagine him doing a nice demonstration cutting up a long beam and assembling the first few sets of cubes so that everyone could see what was going to happen if he kept on going forever.

A much simpler example that makes the same point about infinities was suggested by Galileo. It shows his familiarity with the medieval fascination with infinity and sharpens our appreciation of the central paradox. It is interesting that Galileo raises the matter in his book of Dialogues, which was a work of 'popular' science for all literate people to read. It presented important ideas and discoveries in dramatic form using dialogues and argument to bring out the truth of his ideas.

GALILEO'S PARADOX

'You can tell whether a man is clever by his answers. You can tell whether a man is wise by his questions.'

Naguib Mahfouz[4]

Here in the imaginary dialogue that Galileo[5] created we find written down in its simplest form the key paradox about infinite collections that distinguishes them from finite ones. Galileo knows there is something mysterious about infinity, as did Albert of Saxony, but like Albert he makes no attempt to resolve the puzzle. Galileo reveals[6] these mysterious properties one by one. The dialogue is shown overleaf.

Fig 4.2
Albert of Saxony (1316–90).[7]

A Dialogue *between Salviati, Sagredo and Simplicio:*

Sag: I take it for granted that you know which of the numbers are squares and which are not.
Sim: I am quite aware that a squared number is one which results from the multiplication of another number by itself; thus 4, 9, etc., are squared numbers which come from multiplying 2, 3, etc., by themselves.
Salv: Very well; and you also know that just as the products are called squares so the factors are called sides or roots; while on the other hand those numbers which do not consist of two equal factors are not squares. Therefore if I assert that all numbers, including both squares and non-squares, are more than the squares alone, I shall speak the truth, shall I not?
Sim: Most certainly.
Salv: If I should ask further how many squares there are one might reply truly that there are as many as the corresponding number of roots, since every square has its own root and every root its own square, while no square has more than one root and no root more than one square.
Sim: Precisely so.
Salv: But if I inquire how many roots there are, it cannot be denied that there are as many as there are numbers because every number is the root of some square. This being granted we must say that there are as many squares as there are numbers because they are just as numerous as their roots, and all the numbers are roots. Yet at the outset we said there are many more numbers than squares, since the larger portion of them are not squares. Not only so, but the proportionate number of squares diminishes as we pass to larger numbers. Thus up to 100 we have 10 squares, that is the squares constitute 1/10 part of all the numbers; up to 10,000, we find only

1/100 part to be squares; and up to a million only 1/1000 part; on the other hand in an infinite number, if one could conceive of such a thing, we would be forced to admit that there are as many squares as there are numbers all taken together.

Sag: What then must one conclude under these circumstances?

Salv: So far as I see we can only infer that the totality of all numbers is infinite, that the number of squares is infinite, and that the number of their roots is infinite; neither is the number of squares less than the totality of all numbers, nor the latter greater than the former; and finally the attributes 'equal', 'greater', and 'less', are not applicable to infinite, but only to finite, quantities . . . I answer him that one . . . does not contain more or less or just as many points as another, but that each . . . contains an infinite number . . . So much for the first difficulty.

Sag: Pray stop a moment and let me add to what has already been said an idea which just occurs to me. If the preceding be true, it seems to me impossible to say either that one infinite number is greater than another or even that it is greater than a finite number, because if the infinite number were greater than, say, a million it would follow that on passing from the million to higher and higher numbers we would be approaching the infinite; but this is not so; on the contrary, the larger the number to which we pass, the more we recede from [this property] of infinity, because the greater the numbers the fewer [relatively] are the squares contained in them; but the squares in infinity cannot be less than the totality of all the numbers as we have just agreed; hence the approach to greater and greater numbers means a departure from infinity.

First, Galileo points out that if we list all the positive whole numbers

1, 2, 3, 4, 5, 6, 7 . . . and so on

then this list is infinite, because there is no end to it. If you doubt this then name the last number in the sequence (call it B) and I will be able to produce a bigger number (B + 1) by adding 1 to it.

Now, if we square every number in the list by multiplying it by itself, then to each integer there corresponds one square number:

$$1 \times 1 = 1,\ 2 \times 2 = 4,\ 3 \times 3 = 9,\ 4 \times 4 = 16,\ 5 \times 5 = 25, \text{ etc.}$$

The list of squared numbers (1, 4, 9, 16, 25, . . .) is therefore also infinite because it is in one-to-one correspondence with the infinite list of integers. Think of there being a string tied between each number and its square.

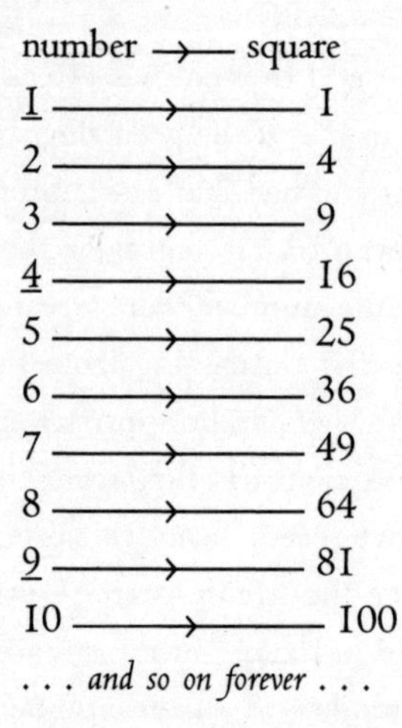

We now have two lists shown in the columns above.

Now, Galileo asks, which list is bigger? Every entry in the list of squares is tied to one and only one entry in the list of integers so it looks as if they must be equally numerous – the same size. But there is a paradox. Every entry in the list of squares (right-hand column) will also occur somewhere in the left-hand column of integers (the first three are underlined in the table above) so surely the left-hand list must be bigger than the right-hand list because it contains lots of other numbers as well!

Galileo did not resolve this paradox. He concludes only that,

> 'we cannot speak of infinite quantities as being the one greater or less than or equal to another'

In fact, Galileo was making slightly heavy weather of his example. There was no need to make his readers struggle with squared numbers. Just think about the correspondence between all the whole numbers (1, 2, 3, 4, . . .) and all the even numbers (2, 4, 6, 8 . . .) that result from doubling them: so 1 links to 2, 2 links to 4, 3 links to 6, 4 links to 8, and so on. Like cups and saucers in a tea set they are paired one to one. Again, there is a unique one-to-one correspondence between the infinite list of numbers and the infinite list of even numbers. Yet all the even numbers are contained in the first list despite the fact that 'common sense' says there must only be half as many even numbers as there are whole numbers!

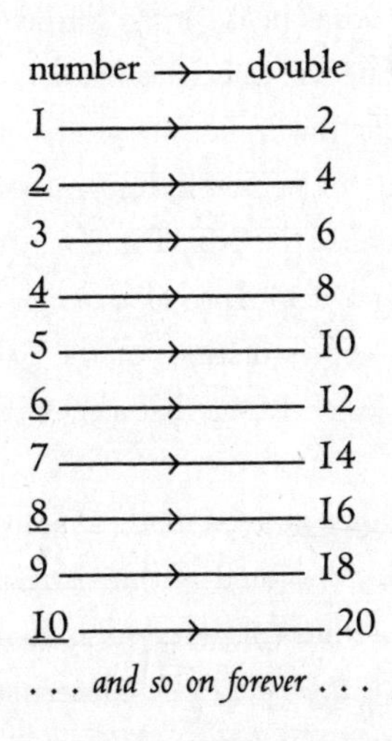

The important thing to appreciate about these examples is that they reveal something unique to infinite collections. If we had taken *finite* lists of things, then they can only be put in one-to-one linkage with each other if they contain an *equal* number of things. For instance, a finite list of married couples contains an equal number of males and females.

Bob ⟶ Jill
Jim ⟶ Joyce
Ron ⟶ Louise
Roy ⟶ Carol

What Galileo's paradox reveals is that infinite collections are not like this: they seem to be able to contain themselves as subsets with plenty left over!

There is a temporal counterpart of these 'getting something for nothing' paradoxes that is usually called the Paradox of Tristram Shandy. It takes Tristram Shandy a whole year to complete an account in his diary of one day in his life. He completes his entry for 1 January, 1760, at midnight on 31 December, 1760; his entry for 2 January, 1760, at midnight on 31 December, 1761, and so on. All the time he is getting further and further behind. If he lives for a finite time he will only have written diary entries for a fraction of the days of his life. But if he lives forever there will be no day of his life for which he has not written a diary entry.

There is also a spatial counterpart, dubbed the 'Map Paradox', which arises when you begin to think of making a map that has a one-to-one scale. We are used to maps that are partial representations of the Earth's surface but, as the American philosopher Josiah Royce first suggested,

> 'suppose that this our resemblance is to be made absolutely exact . . . A map of England, contained within England, . . . One who, with absolute exactness of perception, looked down upon the ideal map thus supposed to be constructed, would see lying upon the surface of England, and at a

definite place thereon, a representation of England on as large or small a scale as you please . . . This representation, which would repeat in the outer portions the details of the former, but upon a smaller space, would be seen to contain another England. And this another, and so on without limit.'[8]

This paradox has been frequently revisited by everyone from Lewis Carroll to Jorge Luis Borges. It is essentially a self-reference paradox rather than a paradox of the infinite. To bring the infinite possibilities into play one could instead stand between two plane-parallel mirrors and look at the never-ending line of reflections of reflections (Figure 4.3) that stretch out like the ghosts of Banquo to the crack of doom.

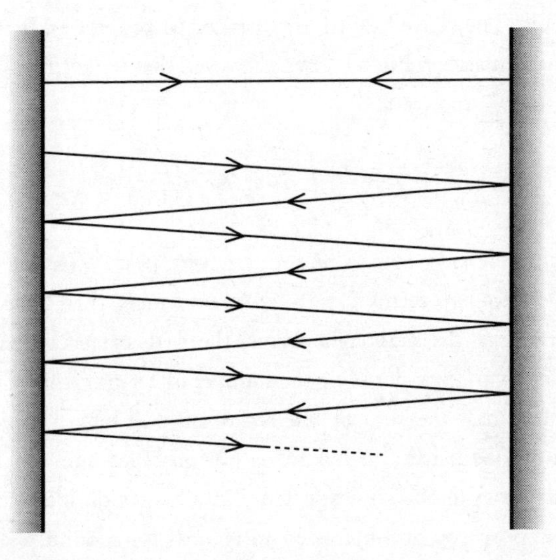

Fig 4.3 *Two parallel mirrors produce a seemingly infinite number of self-reflections. In practice the number is finite because the silvering of the mirrors is not perfect and light is scattering if it moves in a medium that is not a perfect vacuum. Light moves with a finite speed, and so even in perfect conditions an infinite number of images would need an infinite number of reflections to occur and would take an infinite time to produce.*

In reality there are only a finite number of images. The reflectivity is not perfect and the atmosphere scatters light out of the beam's path. Yet, the effects are striking and provide us with one of the simplest and closest snapshots of a potential infinity.

CADMUS AND HARMONIA

'Prove all things.'

St Paul[9]

Mathematicians have long been enchanted by never-ending sequences of numbers. They have beautifully unexpected properties. In 1350 the French mathematician Nicole Oresme proved that the infinite harmonic series of decreasing terms

$$1/1 + 1/2 + 1/3 + 1/4 + 1/5 + 1/6 + 1/7 + 1/8 + \ldots$$

has an infinitely large sum. The proof is very neat. After the first two terms, the next two terms ($1/3 + 1/4$) sum to more than $1/2$, so do the next four terms, the next eight terms, the next sixteen terms, and so on respectively forever, doubling the number of terms gathered together. The result is that the sum of the series must be bigger than an infinite sum of one halves.[10] And this is obviously infinite![11]

This series appears unexpectedly in all sorts of interesting situations. Suppose you are interested in records for natural phenomena, like record annual rainfalls or high tides.[12] In year 1 of keeping records the rainfall will have to be a record. In year 2 the rainfall has a $1/2$ chance of being a record – if it is greater than that of year 1. The expected number of record rainfall years in the first two is therefore $1 + 1/2$. Carrying on, we see that there is a $1/3$ chance that year 3 has higher rainfall than years 1 and 2. Keep on going and we see that the expected number of record rainfalls in the first N years of record keeping is

$$1 + \frac{1}{2} + \frac{1}{3} + \frac{1}{4} + \frac{1}{5} + \ldots + \frac{1}{N}$$

To find the number of records expected per century if conditions are random just put N = 100 and add up the terms. The answer is 5.19. Certainly in the United Kingdom at the moment there are a lot more record rainfall years – and other climatic records – than the 5 per century that this simple harmonic series predicts. This implies that the weather variations are not random, and that there is a systematic change underlying the observed variations, similar to that expected from so called 'global warming'. Notice that the infinite value of the sum of the series reflects the intuitive fact that there is always a chance of a new record in an infinite sequence of observations.

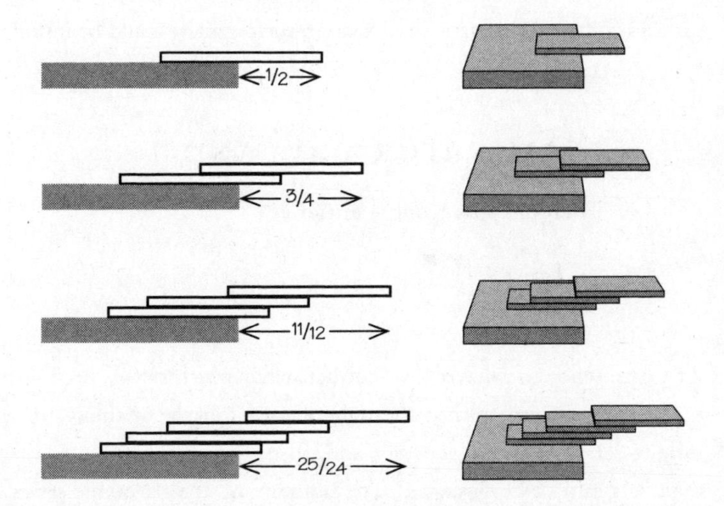

Fig 4.4 *A never-ending stack of books. An infinite number of books can be supported so long as the centre of gravity of the stack never lies beyond the edge of the bottom book. This is possible in principle, not in practice.*

Another nice example of the harmonic series is the book-stacking problem. Pile books on top of one another so that they overhang the side of the table, as shown in Figure 4.4. How far can they protrude over the edge of the table without falling?[13]

They have to be stacked so that the centre of gravity of the stack never lies beyond the edge of the table. Once it does they will start to topple. If each book has size 1 then the maximum possible overhang of N books is just one half of the sum of the harmonic series up to N terms:

$$\text{Maximum Overhang Distance} = \tfrac{1}{2} \times \{1 + \tfrac{1}{2} + \tfrac{1}{3} + \tfrac{1}{4} + \tfrac{1}{5} + \ldots + \tfrac{1}{N}\}$$

The amazing thing about this is that the overhang distance can be made as large as you like by making N big enough.[14] To make the overhang greater than 10 times the size of a single book would need a stack of 272,400,600 books. In an ideal world without friction and imperfect surfaces and smallest particles of matter, the overhang could be infinite!

TERMINATOR 0, ½, AND 1

'There may be trouble ahead.'

Irving Berlin[15]

The harmonic series has a clear-cut behaviour which reveals itself very easily after you have picked the right way of looking at things. If an infinite series of terms does not add up to less than a finite number then it is said to be divergent. The ubiquity of the harmonic series[16] seems to endow it with a harmless familiarity. This is a little misleading. A few examples can restore the apoplexy with which divergent series were so long regarded.

Begin with a simple infinite series, which we suppose is equal to S. It consists of alternating plus and minus 1s, so

$$S = 1 - 1 + 1 - 1 + 1 - 1 + 1 - 1 + 1 \ldots$$

We want to evaluate the sum of this never-ending series. If we first group the numbers appearing in S in pairs as shown below in brackets, the sum of the series is 'obviously' zero, because each bracketed pair of +1 and –1 sums to zero:

$$S = (1 - 1) + (1 - 1) + (1 - 1) + (1 - 1) + \ldots$$

$$S \quad = \quad 0 \quad + \quad 0 \quad + \quad 0 \quad + \quad 0 \quad + \quad \ldots$$

And so the sum is $S = 0$.

But, we could have grouped the terms in the series differently, say by bracketing the next pair along to the right; then

$$S = 1 + (-1 + 1) + (-1 + 1) + (-1 + 1) + \ldots$$

Now we can show that $S = 1$ because each of the bracketed pairs again makes zero, so

$$S \quad = \quad 1 \quad + \quad 0 \quad + \quad 0 \quad + \quad 0 \quad + \quad \ldots$$

So we have proved that $S = 0$ and $S = 1$ and so $0 = 1$!

But why stop there? We can put the brackets down a third way so that

$$S = 1 - (1 - 1 + 1 - 1 + 1 - \ldots)$$

But the unending series in the brackets is just S again, so we have

$$S = 1 - S$$

and so 2S = 1 and S must be equal to 1/2 this time. Armed with these results it is not too great a challenge to show that S can be 'proved' to be equal to any number you like. No wonder the great nineteenth-century mathematicians like Niels Abel[17] avoided divergent series like the plague.

Seen like this, infinities seem to provide the basis for no end of financial scams. We are used to badly behaved computer systems being fixed by simply switching the computer off and back on (this action never fixes my car though). Here we seem to be offered the opportunity of doubling our money just by counting it in a different order. Of course, we know that when our series of alternating 'ones' has only a finite number of terms in it there is no problem at all. Its sum must

Fig 4.5 *Georg Cantor (1845–1918) with his wife, Vally.*[18]

either equal 0 or +1. It doesn't matter how we add up the terms that appear, or where we draw in the brackets, the sum is 0 if there are an even number of terms in the series and +1 otherwise. It is only when you are infinitely rich that your assets depend totally on the order in which you count them up.

Not surprisingly, arguments like this made mathematicians very nervous about infinities. It is easy to see why infinity was regarded as a form of logical plague that destroyed the reliability of everything it touched. In the one subject where infinities could be manipulated clearly they led to disaster. As a result, the desire to banish infinities to some quarantined area away from the rest of logical argument, or to regard them as non-existent, was very strong. At most times in the history of human thought there have been mathematicians who wanted to rid their subject of them except as a form of shorthand for sums of things that have no end.

Out of all this ambiguity and confusion, clarity emerged suddenly in the nineteenth century, due to the single-handed efforts of one brilliant man. Georg Cantor (1845–1918) produced a theory that answered all the objections of his predecessors and revealed the unexpected richness hiding in the realm of the infinite (Figure 4.5). Quite suddenly actual infinities became part of mathematics – but not without a struggle.

COUNTABLE INFINITIES

> 'That action is best which procures the greatest happiness for the greatest numbers.'
>
> Francis Hutcheson[19]

Cantor took the paradoxes that were anathema to mathematicians and used them as the basis for a clear understanding of infinities.

Realising the crucial significance of the strange paradoxes of Albert and Galileo, he changed their status from ill-fitting cast-offs to the central cornerstone of a new theory. Cantor *defined* a countable infinity to be one that can be put into one-to-one correspondence with the list of natural numbers 1, 2, 3, 4, 5, 6, . . . So, for example, the even numbers are countably infinite, so are all the odd numbers. Here is the correspondence for the first nine odd numbers.

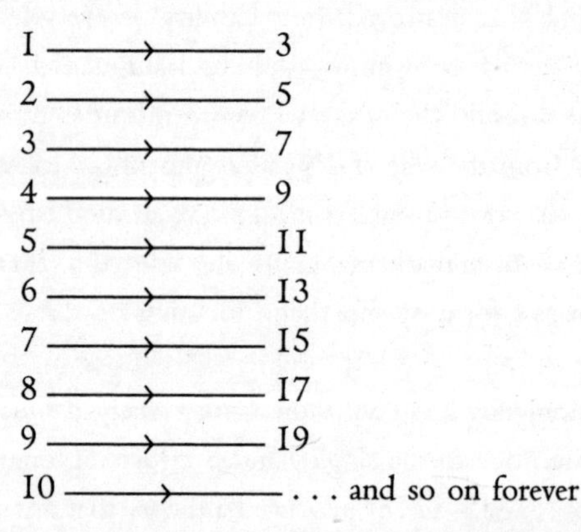

All countably infinite sets therefore have the same 'size' in Cantor's sense. Cantor thought that they were the smallest infinities that could exist and so he denoted them by the first letter of the Hebrew alphabet, the symbol Aleph-nought, $\aleph_0$. Notice how this definition excludes any finite set of objects. Like your tea set of cups and saucers, a finite set can only be put into one-to-one correspondence with another which contains the same number of members (one cup for one saucer).

This leads to some surprising conclusions. Cantor showed that all the fractions formed by dividing one whole number by another (for example $2/3$ or $11/12$) are also countably infinite. The trick is to find a system for counting them so that none get missed out. He used a famous diagonal picture to do this. It counts them row by row in the following order:

$^1/_1$,
$^2/_1$, $^1/_2$,
$^1/_3$, $^2/_2$, $^3/_1$,
$^4/_1$, $^3/_2$, $^2/_3$, $^1/_4$,
$^1/_5$, $^2/_4$, $^3/_3$, $^4/_2$, $^5/_1$,
$^6/_1$, $^5/_2$, $^4/_3$, $^3/_4$, $^2/_5$, $^1/_6$, . . .
and so on forever.

The trick is that along each row the numbers on the top and bottom of each fraction add up to give the same number (so in the 4th row down they all add up to 5, i.e. (4+1), (3+2), (2+3), and (1+4)). This creates a definite order for counting all the fractions which will not miss out any one of them. Whereas we might have thought there were vastly more fractions than single numbers, they are equally numerous when counted Cantor's way. All the infinities that were discussed by mathematicians and philosophers in ancient times were countable infinities in Cantor's sense. But are there any others?

UNCOUNTABLE INFINITIES

'Al-Gore-rhythm: a mathematical operation, which if repeated many times, leads to the desired result – especially in Florida.'

Anonymous

Cantor then showed by a new type of mathematical argument that there were bigger, 'uncountable', infinities. The decimals (most of which are never-ending and include the irrational numbers which cannot be written as fractions) could not be counted systematically. They were 'uncountably' infinite. He proved this in a cunning way. Assume that they can be counted. This means that we must be able to draw up all the unending decimals that do not end in an infinite string of zeros[20]. The first few in the list might look like these

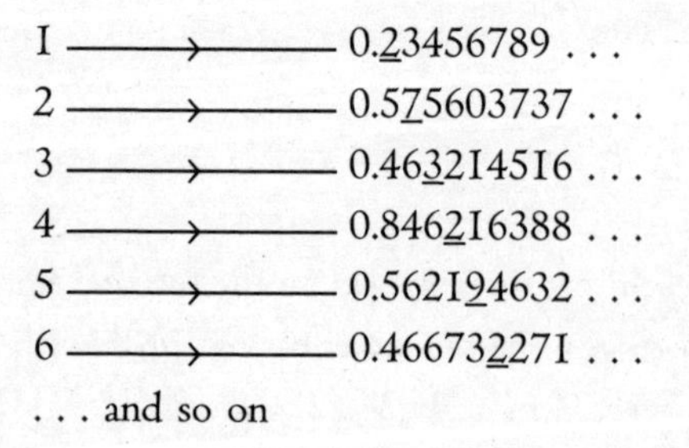

Now we are going to create a new decimal by taking the first digit after the decimal point from the first number, the second digit from the second number, and so on forever. I have underlined the digits we are to use as the digits in our new number. The new decimal begins as follows

0.273292 . . .

Now create a new decimal from this one by adding 1 to every one of its infinitely many digits. We get

0.384303 . . .

The remarkable thing about this number is that it *cannot* appear anywhere on the original ordered listing of all the decimals that we had assumed must exist. It must always disagree with every number in the list by at least one of its digits because it was explicitly constructed like that. Therefore the decimals (sometimes called the real numbers or the continuum of numbers) are uncountably infinite. They are infinitely bigger than the natural numbers or the fractions. Cantor wanted to know if they were as numerous as the next order of infinity, that was denoted by the

Hebrew symbol Aleph-one, $\aleph_1$. Cantor believed that there was no possible infinite collection that was bigger than $\aleph_0$ but smaller than $\aleph_1$, but he was never able to prove it. It turned out to be one of the great problems of mathematics and one that had a most unusual resolution.

This discovery by Cantor – that there are infinities of different sizes and they can be distinguished in a completely unambiguous way – was one of the great discoveries of mathematics. It was also completely counter to the prevailing opinion.

Cantor's predecessor, Bernhard Bolzano (1781–1848), shown in Figure 4.6, began thinking about the paradoxes of the infinite in 1847 when he was sixty-seven years old.

Fig 4.6 *Bernhard Bolzano (1781–1848).*[21]

He came to believe that all infinities were equal. The reason can be seen most simply by looking at another of the 'paradoxes' that Galileo and his medieval predecessors liked to exhibit to challenge the coherence of the idea of the infinite.[22] Take a piece of string and use it to make a semi-circle that has a diameter of one metre. Now imagine an infinitely straight line drawn underneath the semi-circle, parallel to the diameter, see Figure 4.7.

If we draw any straight line from the centre of the semi-circle down to the infinite straight line then it will always cut through the semi-circle at some point on its circumference. The remarkable thing is that the diagram makes it obvious that there is a line like this that links every point on the circumference of the semi-circle to one and only one point on the infinite straight line. So there must be the same number of points on the circumference of the semi-circle as on the line. Moreover, suppose we draw more semi-circles having the same centre but smaller radii. Then the set of all possible straight lines from the centre would pass through every point on the circumference of every circle and each would be in correspondence with every point on the circumference of every other circle. Thus it was argued all these circles contain an infinite number of points on their circumferences and they are all equal in number.

Bolzano concluded that infinite sets are 'equal' because they can

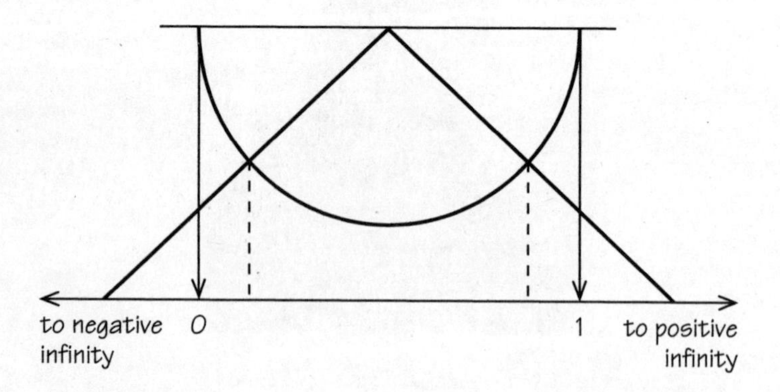

Fig 4.7 *The one-to-one correspondence between a line of one unit in length stretching horizontally from 0 to1 and the entire infinite line from negative infinity* (left) *to positive infinity* (right). *Take any point you choose on the line between negative infinity and positive infinity. Join it by a straight line to the centre of the semi-circle that we have drawn. Where this line cuts the semi-circle we drop a dashed line vertically downwards to pick out a point on the line between 0 and 1. By this process every point on the original line of infinite length ends up at one point on the finite part of the lines between 0 and 1.*

be linked by a correspondence like this. Cantor provided a beautiful example to show that this was not so. Not only were the never-ending decimals – what we called the 'real' numbers – infinitely bigger than the number of whole numbers or fractions, but there could be infinities that were infinitely bigger still.

THE TOWERING INFINITO

'Somebody has to have the last word. Otherwise, every reason can be met with another one and there would be no end to it.'

Albert Camus[23]

Cantor's most dramatic discovery was that infinities are not only uncountable, they are insuperable. He discovered that a never-ending ascending hierarchy of infinities must exist. There is no biggest of all that can contain them all. There is no Universe of universes that we can write down and capture. Before we see how he did that, it is important to say a little about the meaning of the word 'exists' in this context. We are used to using the word on an everyday basis without any ambiguity. 'Cambridge exists', 'inflation exists', seem to be assertions that are clear enough. They are about physical existence. Up until the early nineteenth century, mathematical existence was rather similar. Euclid's geometry existed because it was manifested in the physical world. Indeed, it was believed for thousands of years that there could not be another logically consistent and complete geometrical system. The discovery of non-Euclidean geometries which described the topography of curved surface changed that view. Gradually mathematicians lighted upon a new concept of existence. Mathematical 'existence' meant only logical self-consistency and this neither required nor needed physical existence to complete it. If a mathematician could write down a set of non-contradictory axioms and rules for deducing true statements from them, then those statements would be said to 'exist'. They exist in the same way that positions exist in a game like chess. They are developments according to the rules from the starting position (the axioms). Now it happens that in chess these positions are usually made physical by chess pieces on a fixed board – but this is not necessary. Some experts play in their heads without pieces or a board; others can play by mailing the coordinates of the positions on an imaginary board. So it is with mathematics. Some examples of mathematical existence do have physical existence, but most do not.

When Cantor set about showing that an unending catalogue of mathematical infinities exists, his first aim was to demonstrate mathematical existence: to show that precise definitions of things like infinite sets lead to the conclusions that ever larger ones can be defined. Whether they exist in physical reality is another, quite different, question.

At first you might think that making bigger infinities is child's play. Suppose you have an infinite collection of numbers 1, 2, 3, . . . Just add one more thing to it – say the object *. Isn't that bigger? Unfortunately not; this is just the situation of the Infinite Hotel. Adding one, or two, or even all the whole numbers to a countable infinity still leaves a countable infinity. In Cantor's sense it is the same size. In order to jump up a level to a new order of infinity something different is required, as we saw with the introduction of the never-ending decimals, or 'real' numbers, that are uncountably infinite.

Cantor was able to show that there is no end to the ascending hierarchy of infinities. If you have any infinite set, then you can generate one that is infinitely bigger by considering the set that contains all its subsets. This is called its *power set*. As a finite example consider the set[24] of three objects {A,B,C}. (These could be people and the 'sets' groups of friends, families, or secret societies.) It contains subsets containing the following members (conventionally we include Ø, the empty set which has no members, and the complete set itself in the list of subsets):

$$\{\varnothing\}, \{A\}, \{B\}, \{C\}, \{A,B\}, \{A,C\}, \{B,C\}, \{A,B,C\}$$

There are $8 = 2 \times 2 \times 2 = 2^3$ subsets. In general, if the original set has N members then there are $2^N = 2 \times 2 \times 2 \times 2 \times \ldots$ (N times) possible subsets and members of its power set.

Thus from an infinite set like $\aleph_0$ we can create an infinitely larger set (by which we mean one that cannot be put in one-to-one correspondence with it) by forming its power set, P[$\aleph_0$]. Now we

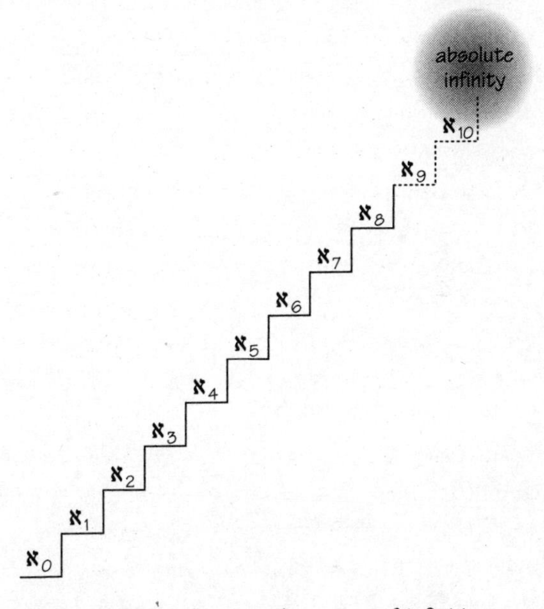

Fig 4.8 *The topless ascending tower of infinities.*

can do the same again by forming the power set of P[$\aleph_0$]. That will be infinitely bigger than P[$\aleph_0$]. And so on, without end.

Thus mathematics gives a never-ending hierarchy of ascending infinities (Figure 4.8). Infinity can never be captured by formulas. This is reminiscent of the ancient attempts to articulate the unreachable nature of God and the Infinite that are found in the great theological writers of the past. It also shows that the number of possible truths is infinite.[25]

These ideas had many theological and philosophical consequences and Cantor found that his ideas about the infinite were well received by scholars in these fields. Alas, within mathematics the story was quite different, as we shall see.

chapter five

The Madness of Georg Cantor

'To be listened to is a nearly unique experience for most people. It is enormously stimulating. Man clamors for the freedom to express himself and for knowing that he counts.'

Robert C. Murphy[1]

CANTOR AND SON

'I continued to do arithmetic with my father, passing proudly through fractions to decimals. I eventually arrived at the point where so many cows ate so much grass, and tanks filled with water in so many hours. I found it quite enthralling.'

Agatha Christie[2]

Cantor & Co. was a successful international wholesale business, and as a result young Georg Cantor was one of six children who grew up in comfortable circumstances, attending good private schools in Frankfurt. Georg had many talents and might well have pursued a career as a musician, as did some of his relatives, or as an artist. Yet in his teenage years he became increasing captivated by mathematics, physics, and astronomy. His father, Georg senior, was strongly supportive of all his studies and also imposed his strong religious beliefs in destiny upon his son; some biographers have wondered whether the paternal support was really just a case of the father's own unfulfilled ambitions being pursued through the life of his eldest son. Yet, for all this, Georg junior seems to have survived his life at home, and he graduated from Darmstadt School in 1862, aged seventeen, with high marks, moving first to study mathematics at the Polytechnical Institute in Zurich and then going on to the famed mathematics course at the University of Berlin, the centre of the mathematical world in the mid-nineteenth century. There he encountered great mathematicians like Karl Weierstrass, Sophie Kowalewski, and Ernst Kummer, who followed in the footsteps of men like Bernhard Riemann and Peter Dirichlet. He was also taught by the influential Leopold Kronecker.

Cantor followed the usual route of a young academic of the day, jumping through the hoops laid out for him by completing his degree and then his doctorate in Berlin, before beginning a form of apprenticeship which involved teaching classes of pupils at the university in the city of Halle, a medieval city famous for being the birthplace of the great seventeenth-century composer George Frederick Handel. Halle University was an in-between place for a budding mathematician, geographically half-way between the great universities of Berlin and Göttingen; it was the sort of place that you hoped would be a stepping stone to becoming a professor at one of these two famous mathematical centres.

Unfortunately for Cantor, that call never came and he spent the whole of his career in the minor mathematical department in Halle – where there were few visitors and no mathematicians of Cantor's calibre – living comfortably in a big house with his close family following his marriage in 1875 to his sister's friend, Vally Guttmann. Things were to become more exciting for Cantor, but not in ways that he could have wished.

THE CHRONICLE OF KRONECKER

'Logic sometimes makes monsters'

Henri Poincaré[3]

The year 1871 was a watershed in Cantor's career as a mathematician. Until that time, his former professor in Berlin, Leopold Kronecker, had been on good terms with him, sympathetic to his work and helpful in getting him established in Halle. He even provided some important mathematical suggestions which helped Cantor to complete some of his first research papers. Then something changed. Cantor began to work on infinities, and in Kronecker's eyes he had suddenly become 'a corrupter of youth'.[4]

Kronecker was the son of a wealthy Prussian businessman and was in no need of a university salary to support his mathematical career

Fig 5.1 *Leopold Kronecker (1823–91).*[5]

(Figure 5.1). He did important work on algebra and number theory in Berlin, but had to spend a period of eleven years away from mathematics while running the family business. Eventually, he returned to become a professor in Berlin in 1882.

The historian of mathematics David Burton writes that

> 'Kronecker was a tiny man, who was increasingly self conscious of his size with age. He took any reference to his height as a slur on his intellectual powers. Making loud voice of his opinions, he was venomous and personal in his attacks on those whose mathematics he disapproved; and his opinions relative to the new theory of infinite sets were ones of ire and indignation . . . Kronecker categorically rejected [Cantor's] ideas [about infinite sets] from the start. He asserted dogmatically, "Definitions must contain the means of reaching a decision in a finite number of steps, and existence proofs must be conducted so that the quantity in question can be calculated with any required degree of accuracy."'[6]

Any discussion of infinite sets was, according to Kronecker, illegitimate since it began with the assumption that infinite sets exist in mathematics.

Kronecker wanted to define mathematics to consist only of those deductions that could be made in a finite number of steps from the natural numbers (1, 2, 3, 4 . . .). This goal is encapsulated in a famous remark he made in a speech: 'God created the natural numbers, and all the rest is the work of man.'

Kronecker was not alone in holding such views, but he was the most influential and vociferous advocate of the mathematical straitjacket called 'finitism'. He believed that we should only do mathematics by building up quantities and arguments in a finite number of steps. Today, this would be classed as the mathematics that a computer could carry out if correctly programmed. We know that this is a small fraction of what is allowed to be mathematics if we do not restrict ourselves to finite step-by-step deductions.

Kronecker would not allow you to assume that something exists if you could not explicitly describe how it could be constructed. Likewise, he would not admit into mathematics those proofs which showed that something must exist without giving the step-by-step recipe for arriving at its construction. In effect, Kronecker believed in a smaller scope for mathematics than did most other mathematicians.

Up until the work of Cantor on infinities, it had been possible to take Gauss's view that infinities in mathematics were always *potential* infinities, and so mention of 'infinity' was just a shorthand for describing a series or a process that had no end: you didn't *do* anything with these infinities. You didn't use them to prove other things were true.

Gauss, the greatest mathematician of the day, had set the tone when he wrote in a letter to his friend Schumacher in 1831 that

> 'I protest against the use of infinite magnitude as something completed, which in mathematics is never permissible. Infinity is merely a façon de parler, the real meaning being a limit which certain ratios approach indefinitely near, while others are permitted to increase without restriction.'

In universities all over the continent of Europe, the division between potential and actual infinities was regarded as crucial, and the general view was that only potential infinities were meaningful.

Despite this current of opinion, most mathematicians held mild views on the issue and rarely encountered a problem where taking a view about finitism really mattered. As a result, most were surprised, and many were irritated, by Kronecker's outspoken finitist views – but the highly-strung, increasingly paranoid Cantor was the most seriously affected by Kronecker's criticisms. All of his work was focused upon defining and manipulating actual infinities and Kronecker characterised this work as a study of things that did not exist, and total 'humbug'![7]

Cantor's hopes of becoming professor of mathematics at the University of Berlin were totally blocked by Kronecker's opposition. Kronecker's influence extended far beyond Berlin, and at Göttingen as well Cantor was repeatedly passed over in favour of seemingly less-distinguished candidates. Kronecker also sat on the editorial boards of journals which delayed or prevented the publication of some of Cantor's work. As a result, Cantor spent his entire professional career, forty-four years, at Halle University, a small college with no mathematical reputation.

Yet Cantor did get his important work published between 1874 and 1884, and it was well known, if occasionally controversial, amongst his young colleagues in Germany at the time – all the more reason for his despair about his lack of advancement. Cantor eventually became so angered by Kronecker's attacks that he wrote directly to the Ministry of Education, hoping to annoy Kronecker by applying for a position vacant in Berlin the following spring. He wrote to his old friend Gösta Mittag-Leffler on 30 December 1883, telling of his desperate measure:

> 'I never thought in the least I would actually come to Berlin . . . since I know that for years Schwarz and Kronecker have intrigued terribly against me, in fear that one day I would come to Berlin, I regarded it as my duty to take the initiative and turn to the Minister himself. I knew precisely the immediate effect this would have: that in fact Kronecker would flare up as if stung by a scorpion, and with his reserve troops would strike up such a howl that Berlin would think it had been transported to the sandy deserts of Africa, with its lions, tigers, and hyenas. It seems that I have actually achieved this goal!'[8]

Kronecker responded the following month by himself writing to Mittag-Leffler (the editor of *Acta Mathematica*) asking if he could publish in his journal a short article setting out his views about certain mathematical conceptions in which he would show that 'the results of modern . . . set theory [i.e. Cantor's work] are of no real significance'.[9]

Actually, Kronecker had no intention of publishing such a paper, but simply wanted to rattle Cantor into refusing to publish in Mittag-Leffler's journal again in the belief that the editor had betrayed his faith in him by agreeing to publish Kronecker's paper.

At first, however, Cantor was pleased to hear of Kronecker's intention to write a critical article, as it would make Kronecker's opposition public and he would be able to answer it. But then, as Kronecker hoped, Cantor seems to have become suspicious that it would degenerate into personal polemics and told the editor that if the journal published anything critical from Kronecker, he would not support the journal with any of his own work in the future. Kronecker never did send anything to the journal, and the events show something of Cantor's paranoia and despair.

In 1884 Cantor attempted to cool things down by writing directly to Kronecker in a spirit of reconciliation and they had several discussions. However, although Kronecker was outwardly conciliatory, no real peace was made. Cantor concluded there was little hope of success. Indeed, any success Cantor had with others made Kronecker feel even more threatened by Cantor's ideas. Cantor says that, 'It seems to me of no small account that he and his preconceptions have been turned from the offensive *to the defensive* by the success of my work.'[10]

Soon afterwards Mittag-Leffler suggested that one of Cantor's papers should not be published in his journal, saying diplomatically that its insights were 'one hundred years too soon'. This was devastating to Cantor and he never published in the journal again, saying 'I never want to know anything again about *Acta Mathematica*'. (He had also, in 1878, resolved never to publish again in *Crelle's Journal*, another mathematics journal influenced by Kronecker.) As a result, by 1885 he had decided to give up mathematics entirely (Figure 5.2).

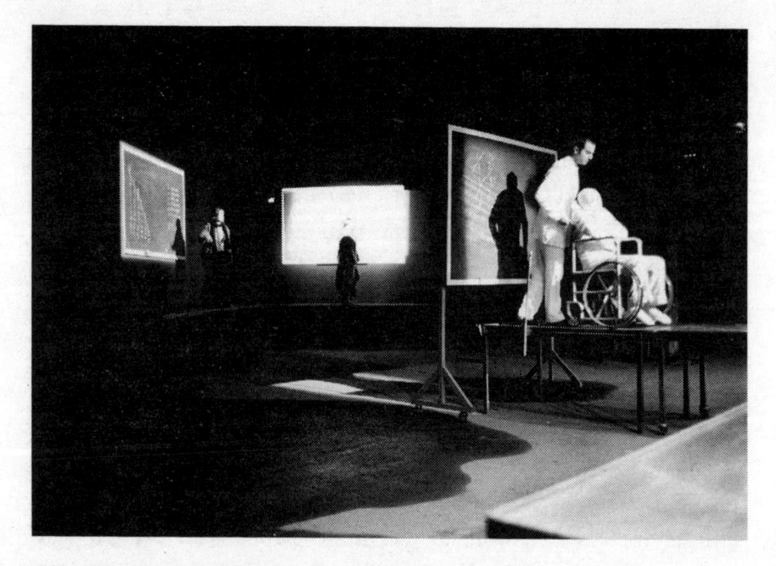

Fig 5.2 *Pictures of Cantor's struggle with mathematics and mathematicians from the Milan production of* Infinities.

Cantor's belief that he was being persecuted by Kronecker because of his mathematical views led to a complete nervous breakdown in 1884. He made a recovery one month later, but subsequently his life was punctuated by bouts of depressive illness which forced him to stay for periods in the clinic in Halle. In the intervals when his mind was clear, he spent a lot of time on studies of the ancient history of counting systems, theology, and history. It was not until the 1900s, when he had finished his research, that Cantor's work started to receive international recognition, with many prizes and honorary degrees being awarded to him. However, this recognition came mainly from outside Germany and Cantor complained, in 1908, of the German mathematicians 'who do not seem to know me, although I have lived and worked among them for fifty-two years'.

Ultimately, as we will see, these events and stresses tipped him into depression and undermined his belief in the worth of his own or any mathematical research. He attempted to transfer from the mathematics department to the philosophy department at his university – a request that was refused. Yet, the university bent over backwards to give him time to rehabilitate himself, and hired temporary lecturers to deputise for him during his periods of illness and absence. To escape these periods of depression Cantor began contemplating the theological implications of his work on the infinite. Its reception by theologians was unexpected.

CANTOR, GOD, AND INFINITY – THE TRINITY WITH AFFINITY

'I entertain no doubts as to the truth of the transfinities, which I have recognised with God's help and which, in their diversity, I have studied for more than twenty years; every year, and almost every day brings me further in this science.'

Georg Cantor[11]

In 1885 Cantor put mathematics to one side and started to correspond with theologians and other intellectuals about infinity. Always someone of strong religious faith, and strongly influenced by his father's forceful beliefs, his attitude towards his work on infinity began to shift in an unusual way. He started to tell his friends that he had not been the inventor of the ideas about infinity that he had published. He was merely a mouthpiece, inspired by God to communicate parts of the mind of God to everyone else. This increased his belief in the truth of his work on infinity, for in his mind it had risen to the elevated status of revealed truth.

Cantor had changed direction at just the right time. The mathematical world in his vicinity may have been under the conservative influence of Kronecker's outspoken views, but when Leo XIII ascended to the papacy in 1878 he brought a liberalisation of the Church's attitudes in many areas. He sought to reconcile science and religion by offering a more enlightened lead from Rome.

This was good news for one Constantin Gutberlet, a priest, philosopher and theologian, and one of Germany's leading neo-Thomists. Gutberlet believed controversially that the human mind could grasp actual infinities and talk meaningfully about them. As a result, he had come under attack from Catholic theologians, but had responded by seizing upon Cantor's mathematical work to argue that it provided clear evidence that the human mind could contemplate the actual infinite. Moreover, if it did so, it would get closer to the true nature of the Divine. The collection of divine thoughts in the mind of an unchanging God, he argued, must comprise a complete and infinite set. This was for him evidence that Cantor's infinities actually existed, and to deny it would require you to give up the infinite and absolute mind of God. High stakes indeed.

Gutberlet's approach is reminiscent of the way in which Euclid's geometry had played an important role in supporting claims that the human mind could have access to matters of ultimate truth. If theologians were challenged by sceptics who argued that ultimate truth was something that transcended the human mind, they could point to Euclid as an example of part of the ultimate truth about the Universe that we have found. In the nineteenth century there would be radical changes to our view of mathematical structures like Euclid's geometry. No longer would it be possible to argue that Euclid's geometry was the one and only logically possible geometry and therefore tells how the world must necessarily be. It was recognised that there can exist other non-Euclidean geometries – infinitely many of them – all logically self-consistent. The fact that they exist mathematically by virtue of being logically self-consistent does not mean that they must exist in physical reality though.

Gutberlet wrote about the vital theological importance of Cantor's work, and entered into correspondence with him over the question of the absolute infinity of God's existence. Cantor was extremely interested in the theological consequences of his ideas, and argued that the higher infinities he had found increased the extent of God's dominion for they had no upper bound: there was no 'biggest' infinity. His never-ending tower of infinities provided a simple answer to the challenge that Gutberlet was facing, that understanding and codifying infinity was reducing the status of God. This might well have been worrying to some, had there been a biggest infinity.

Cantor believed that he could use his knowledge to prevent the Church making grave errors about its doctrines concerning infinity. He thought it was a mission to which he had been called. He declared in a letter to a friend, in 1896, that

> 'From me, Christian philosophy will be offered for the first time the true theory of the infinite.'[12]

He also said,

> 'But now I thank God, the all-wise and the all-good, that He always denied me the fulfilment of this wish [for a position at university either in Göttingen or Berlin], for He

> thereby constrained me, through a deeper penetration into theology, to serve Him and his Holy Roman Catholic Church better than I have been able with my exclusive preoccupation with mathematics.'[13]

Many have felt that Cantor was signalling his despair with all that had gone before and was just turning to a less demanding and controversial activity, away from Kronecker and the rivalries of other mathematicians. However, he interpreted his growing liking for theology and philosophy and his disaffection with mathematics as the work of God. He saw himself as a servant of God who had been given the talent for mathematics in order to be of service to the Church.

He gave up contact with his mathematical friends and was happy about his contacts with Church theologians and philosophers who were interested in his work and thought it significant. Religion renewed his self-confidence and convinced him that his work was important after all, despite the opposition of so many mathematicians. In 1887, Cantor wrote to his colleague Heman of his confidence that he could answer any criticism and overcome any opposition:

> 'My theory stands as firm as a rock; every arrow directed against it will return quickly to its archer. How do I know this? Because I have studied it from all sides for many years; because I have examined all objections which have ever been made against the infinite numbers; and above all, because I have followed its roots, so to speak, to the first infallible cause of all created things.'[14]

Georg Cantor was very interested in how mathematics might reveal the existence of God. In letters to Cardinal Franzelin, he indicated that the infinite, or the 'Absolute', belonged uniquely to God. He believed that it was God who ensured that the hierarchy of transfinite numbers existed, stretching beyond the simplest countable infinities, increasing without limit. Because the largest of these could never be captured by a single formula – from any infinite set it was always possible to make an infinitely larger one – Cantor regarded the transfinite numbers as ascending directly to the Absolute, to the 'true infinity' whose magnitude was an absolute maximum that was incomprehensible to mere human understanding. The Absolute Infinite was beyond human determination, since once it was determined, the Absolute would no longer be regarded as infinite, because it would then necessarily be finite by definition – once determined it could be added and subtracted and manipulated or infinitely increased, just like the lesser infinities.

Thus Cantor seems to think of Absolute Infinity in the way that Archbishop Anselm thought of God in his famous 'ontological' proof of the existence of God, as being that above which no greater could be conceived.

What did Cantor's colleagues think about his ideas on God and infinity? Constantin Gutberlet had studied under Franzelin. He corresponded with Cantor and took his ideas very seriously. At first he was worried that Cantor's work on mathematical infinity challenged the unique, 'absolute infinity' of God's existence. However, Cantor assured him that instead of diminishing the extent of God's dominion, the transfinite numbers actually made it greater. After talking to Gutberlet, Cantor became even more interested in the theological aspects of his own theory on transfinite numbers.

Furthermore, Gutberlet argued that since the mind of God was unchanging, the collection of Divine thoughts must comprise an absolute, infinite, complete closed set, and offered this as direct evidence for the reality of concepts like Cantor's transfinite numbers. Like Pythagoras and Plato, Cantor believed that the numbers (particularly his transfinite numbers) were externally existing realities in the mind of God. They were discovered. They followed God-given laws, and Cantor believed it was possible to prove their existence from God's perfection and power. Indeed, Cantor said, it would have diminished God's power had God only created finite numbers.

Ironically, Cantor's love of the infinite had a distinctly anti-Pythagorean flavour. Pythagoras believed infinity was the destroyer in the Universe, the malevolent annihilator of worlds. If mathematics were a war, then the struggle was between the finite and the infinite. The Pythagoreans became obsessed with the negative aspects of infinity. They believed that the whole numbers closest to one (and therefore the 'most' finite in some sense of being farthest from the infinite) were the most pure of all numbers.

ALL'S SAD THAT ENDS BAD

'Behold the heaven of heavens cannot contain Thee'

The book of Chronicles[15]

Leopold Kronecker died in 1891 without ever becoming involved in a public criticism of Cantor's work. After 1895, a few of Kronecker's old allies opposed Cantor's ideas but, increasingly, the younger mathematicians supported Cantor and the dispute over finitism just faded away.[16] Cantor, however, never regained his mathematical powers and his decline had a terrible inevitability about it.

As we have seen, he had suffered his first breakdown in May 1884, just after his thirty-ninth birthday. He returned to doing mathematics in the autumn, but his interests had changed. He spent a lot of time working on Elizabethan history (trying to prove that Francis Bacon wrote Shakespeare's plays!), and early theology.

Eventually he suffered further breakdowns, and was in hospital for part of 1899 because of mental instability. He applied for leave of absence from teaching at Halle and wrote to the Ministry of Culture saying he wanted to leave his professorship. If they would pay him the same salary, he would be happy to take a quiet position in a library somewhere. He wanted to break away from maths and stressed his knowledge of history and theology. He even threatened to apply to join the Russian diplomatic service. All this came to nothing.

In December 1899, while he was out giving a lecture in Leipzig about the Bacon-Shakespeare authorship issue, his youngest son, Rudolf, died suddenly just before his thirteenth birthday. Rudolf, although always frail and in poor health, had been a gifted musician, just as his father had been as a child before he gave up music for mathematics. Despite this cruel blow, Cantor managed to remain of sound mind for three years, but was back in hospital, relieved of his teaching duties again, in the winter of 1902–3. Some of his work was questioned in a public conference in 1904 and this agitated him greatly. He was in hospital during the winter of 1904–5, in 1907–8 and 1911–12. In 1915 an international meeting was planned to celebrate his seventieth birthday, but the war prevented all but a few close German friends from attending. He was admitted to the Halle clinic for the final time on 11 May 1917. He didn't return home. In wartime rationing conditions, food was scarce and he lost weight steadily. He died of heart failure on 6 January 1918, twenty-seven years after Kronecker. At the end of the game, the pawn and the king go back in the same box.

chapter six

Infinity Comes in Three Flavours

'It is incumbent on the person who specialises in physics to discuss the infinite, and to inquire whether there is such a thing or not, and, if there is, what it is.'

Aristotle[1]

TRIPLE TOP

'Why do buses always come in threes?'

Rob Easterway and Jeremy Wyndham[2]

Cantor could build up a never-ending tower of larger and larger infinities *from below*, but he realised that that infinity could not be approached 'from above'. There was no God's-eye view of the tower that was available to us. Cantor used the name *Absolute Infinity* for the totality of everything. It is something that is beyond mathematical determination or representation. It can only be comprehended by the mind of God. Cantor's distinction between the transfinite numbers he had constructed (*mathematical infinities*), infinities in the physical Universe (*physical infinities*), and *absolute infinity* was of profound importance to him. He wrote that

'The actual infinite arises in three contexts: *first* when it is realized in the most complete form, in a fully independent

> other-worldly being, *in Deo*, where I call it the Absolute Infinite or simply Absolute; *second* when it occurs in the contingent, created world; *third* when the mind grasps it *in abstracto* as a mathematical magnitude, number, or order type. I wish to make a sharp contrast between the Absolute and what I call the Transfinite, that is the actual infinities of the last two sorts, which are clearly limited, subject to further increase, and thus related to the finite.'[3]

While Cantor distinguished three levels of infinity: in the mind of God (Absolute), in the mind of man (mathematical) and in the physical Universe[4] (physical), he maintains that God instilled the concept of number, both finite and infinite, in the human mind to reflect His own perfection.[5] He opposed completely the idea that transfinite numbers were merely the mind's invention or some type of mental category to deal with notions that we could not perfectly capture.

It is possible to hold any one of eight alternative views about the existence or non-existence of the three sorts of infinity (mathematical, physical, Absolute). Here are the eight permutations of belief, each with the name of a well-known philosopher or mathematician who appears to have held that combination:[6]

	Mathematical ∞	**Physical ∞**	**Absolute ∞**
Abraham Robinson	No	No	No
Plato	No	Yes	No
Thomas Aquinas	No	No	Yes
Luitzen Brouwer	No	Yes	Yes
David Hilbert	Yes	No	No
Bertrand Russell	Yes	Yes	No
Kurt Gödel	Yes	No	Yes
Georg Cantor	Yes	Yes	Yes

LET'S GET PHYSICAL

'Singularity is invariably a clue'

Arthur Conan Doyle[7]

Infinity comes in three flavours and it is all very well believing in none or more of these varieties of infinity but what is the evidence for your belief? Cantor and his predecessors have established a clear description of *mathematical* infinity – or, rather, infinities, as we have seen that there is a never-ending tower of infinities of ever greater size: none can be called the biggest. Some people, like Kronecker and his disciples, were once unhappy to admit such quantities into mathematics. Today, treating infinite quantities mathematically as though they are actual infinities is a practice that has stood the test of time and is now regarded as an important part of mathematics. If you wish, you can define a smaller mathematics that only uses a finite number of deductive steps, like a computer. This is perfectly consistent and of some interest to logicians, but most applied mathematicians would regard it as unnecessarily restrictive and rather like fighting with one arm tied behind your back. Consequently, you would have to search quite hard today to find someone who is a finitist in Kronecker's sense, and who therefore does not accept that actual mathematical infinities can exist. They are not regarded as a logical timebomb within mathematics.

The second flavour of infinity is what we might call *physical* infinity. A physical infinity is much more dramatic than a mathematical infinity. The latter just happens on paper, but the former could destroy something of the fabric of the universe. Up until the early twentieth century, the laws of Nature that physicists worked with all had a similar pattern. They assumed the existence of an unchanging three-dimensional space and a steady flow of time. Then they provided rules for how things would move around and interact with one another in this three-dimensional space as time passed. Sometimes, rather dramatic things would happen, but, no matter how extreme events became, they never affected the nature of space or the flow of time.

Einstein's insights into the links between space and time and the force of gravity changed this Newtonian conception of things in an important way. In Einstein's development of Newton's ideas, the geometry of space and the rate of flow of time are not preordained and fixed independently of the matter and motion that occur. When you put things down in Einstein's space, then their mass and motion determine the shape of the space and the rate of flow of time at different places (see Figure 6.1). When you are far from the masses the space is almost flat and unaffected by their presence, just like in Newton's picture. However, when large amounts of mass are compressed into a small volume of space and objects move at speeds close to that of light, the distortions of space and time are considerable. In this picture of space and time it is clear that if any physical quantity, like a density or a temperature or an acceleration, were to develop an infinite value, there would be wider repercussions. The curvature of space would be

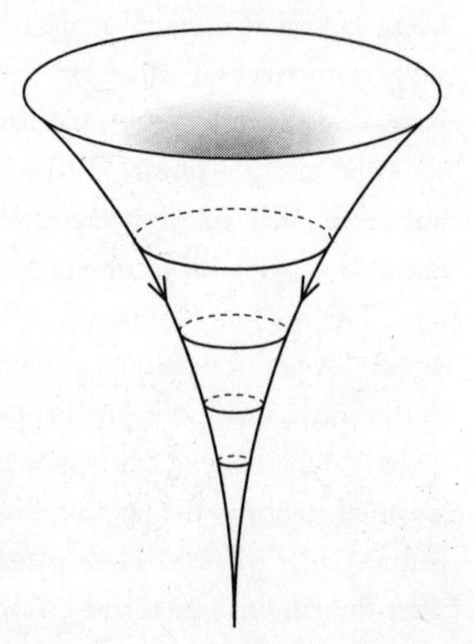

Fig 6.1 *Curved space which develops into a singularity.*

made infinite as well: in effect the space would be torn apart. This type of extreme situation, in which any physical quantity becomes infinite, is called a *singularity*.

Could such a singularity exist in the Universe? Could we witness a physical infinity? This is a big question that has challenged physicists and astronomers for an answer for nearly fifty years. There are different positions on this issue according to the type of scientist you ask. Here are the three principal protagonists:

Engineers meet the infinite

If you are an engineer studying how water flows along a channel or sound travels through air, you will be familiar with the development of a shock wave. The simplest equations you could write down to describe how a fast-moving sound wave develops will predict that a physical infinity occurs when the wave begins to move faster than the speed of sound in the medium in which it is moving (Figure 6.2). For

Fig 6.2 *A moving wave steepens to form a shock.*[8]

air this critical speed is about 750 mph. In practice this infinity never actually occurs. There is a very fast change in the wave that manifests itself as a shock, or sonic boom. In order to conserve energy, the wave has to make a sudden change, rather than continue with a steady increase. We are familiar with this when a supersonic jet, like the Concorde, flies past or there is a clap of thunder, or someone lets off a fire cracker, or a whip is cracked – the tip of the whip moves faster than the speed of sound.

This means that in practice the appearance of apparent physical infinities in the equations describing fluids and aerodynamics is not taken seriously as a manifestation of the infinite. It is just a signal that the modelling of the events under study is incomplete. If more detail is added – including air friction, the stickiness of a liquid or the finite sizes of molecules – then the infinite changes will be smoothed out into large but finite ones.

Experience of this type will make you doubt the reality of any physical infinities and suggests that their appearance might always be down to the failing of human knowledge: a signal that more accurate laws of Nature are required.

Other idealised situations also fail to deliver in reality the infinities that they promise on paper. Think back to the example given earlier, of standing in between two parallel mirrors at the fairground. There should in principle be an infinite number of images of you bounced back and forth between the two mirrors. In reality, the mirrors are not perfectly silvered and the light is gradually degraded by the mirror surfaces and by the air in between them. Even if all was perfect in this respect – perfect reflection and a perfect vacuum between – light still travels with a finite speed and it would take an infinite time for an infinite number of reflections to occur.

Elementary-particle physicists meet the infinite

If you are a particle physicist trying to understand the most basic laws of Nature and the smallest entities possessing mass and energy they govern, then you will also have encountered infinities, lots of them. For nearly fifty years, particle physicists have been used to the appearance of infinite quantities in the mathematical calculations of straightforward physical quantities, like the rate of particular processes in which a particle decays into others. So ubiquitous was this problem that it became known as the 'problem of infinities', and the quest to solve it was a major enterprise in the research agenda. In many ways the search for its solution directed the way in which the subject developed and the standards by which its achievements were judged. Eventually, as a stop-gap, the problem was quenched rather than solved. A systematic method for splitting the answer obtained in a calculation into an infinite and a finite part was arrived at. The infinite part was subtracted, to leave the finite piece that could be compared with what was observed. This mysterious process, dubbed 'renormalisation', produced results of astonishing accuracy – agreement between theory and observation to 16 places of decimals, the most accurate predictions in the whole of human experience. It suggested that these infinities were artefacts of a clumsy way of looking at things.

Since the early 1980s, string theory has shown how these infinities can be avoided by changing our conception of what the most elementary pieces of the Universe are like. The infinities derived from our picture of the most elementary particles of matter as 'points' of zero size that trace out lines when they move in time through space; string theory proposes that they are in fact little loops of energy that trace out tubes in space when they move. These loops have a tension, analogous to an elastic band, that decreases as the temperature of the environment rises to very high values but increases as it falls to the energies in the Universe today. Thus at low energies the tension makes the loops more and more pointlike – and so the picture of Nature's elementary particles as points can be an extremely good approximation to reality, as we have found, but at very high energies it will fail. However, the picture of loops of energy interacting to produce new loops is a smooth process that gives rise to none of the nasty infinities that appear in the pointlike picture. The infinities disappear and all is finite (see Figure 6.3).

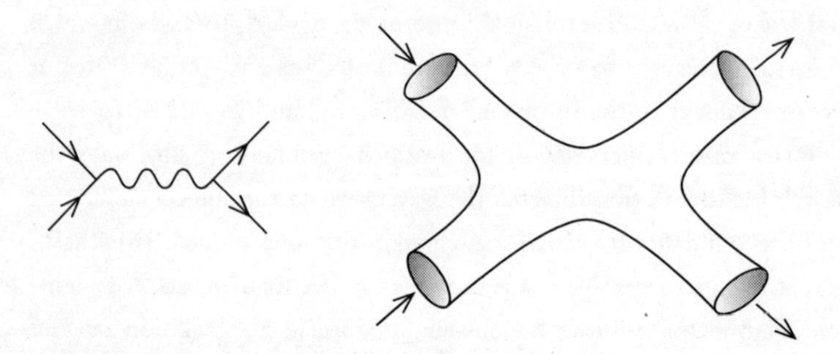

Fig 6.3 *Two moving points trace out lines. Their interaction creates a sharp corner in the picture that follows their motion through space and time. This is the signature of an infinity in calculating the nature of this interaction. By contrast, two moving loops trace out tubes when they move and their interaction is described by a smooth transition to two other tubes with no sharp corners. The seamless double pair of trousers that results is the signature of a process that contains no hidden infinities.*

While we don't yet know if string theory is the true theory of matter and energy at its most basic level, the development and reception of this theory by particle physicists reveals their true feelings about physical infinities. They don't believe in them! As in the study of fluids, if an infinity appears in a calculation about elementary particles, it is regarded as a defect of the theory, showing it to be an approximation that is outliving its usefulness. It is invariably believed that a bigger and better theory will always exorcise the infinities.

Cosmologists meet the infinite

If you are a cosmologist then the problem of physical infinities is more complicated and diverse than it is for particle physicists and engineers. Infinities can appear in all sorts of different ways with different strengths. Some possible cosmological infinities are clearly of the 'potential' sort. If the Universe has an infinite size and infinite future lifetime, then these might represent actual infinities from the superhuman point of view of someone looking at the Universe from outside space and time, but for us they are never actual infinities.[9] One must also worry about what quantities are 'physical'. It is easy to define a quantity which takes on an infinite value somewhere in the Universe, but that does not mean that we will be able to measure or experience it.

We can never know by direct observation that the Universe is infinite in size, rather than merely finite but enormously large.[10] However, the most important questions about the reality of physical infinities in the Universe are much more concrete. Can there be places in the Universe where measurable physical quantities like the density or the temperature

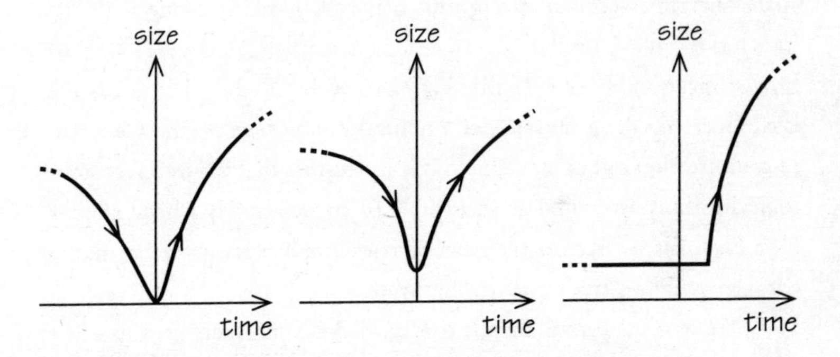

Fig 6.4 *Three possible beginnings for the observed expansion of the universe: about 13.7 billion years ago the expansion could have been preceded by a 'crunch' of infinite density, a gentler 'bounce' at finite density and temperature which resulted from a previous state of contraction, or by a non-expanding state that suddenly bursts into motion.*

of matter can become infinite? Could anything of finite extent be crushed to zero size and infinite density in a finite time from now?

The response is quite divided. There are those, like the particle physicists and engineers, who see the prediction from Einstein's equations that the Universe can have had a beginning in a state of infinite density as a signal that Einstein's equations cease to apply when the density of matter gets too high. They believe that the search for an improved theory will render these infinities finite as well. Instead of a crunch in the past there was a gentle 'bounce' or a coasting phase of the Universe's expansion (See Figure 6.4).

There is good reason to think like this. Einstein's theory may just be a low-energy approximation to string theory when the string tension gets high, and string theory has already shown that it can get rid of all sorts of other infinities. Maybe it can rid us of those at the start of the Universe as well? This is the hope of Stephen Hawking, who sees the infinities of Einstein's theory of gravity as a signal that a quantum theory of gravity is needed to supersede it.

Many people see the infinite beginning of the Universe, where space and time seem to spring into being ready-made, along with the impetus for the Universe to expand, as a mathematical expression of Divine creation. In 1952, the Vatican embraced the picture of the expanding Big Bang universe as a natural conception of the Christian idea of creation out of nothing.[11] It is interesting that the initial cosmological infinity is treated as acceptable by many scientists because they have been made used to the idea of the Universe having a beginning through religious traditions in the West.

Yet it is dangerous to put too much faith in events at a moment where the density of the Universe is infinite. As Stephen Hawking advises, regarding the deduction of an infinity at the beginning of the Universe:

> 'Although many people welcomed this conclusion, it has always profoundly disturbed me. If the laws of physics could break down at the beginning of the universe, why

> couldn't they break down anywhere? . . . predictability would completely disappear.'[12]

Long ago, Einstein himself took a rather similar negative attitude to the appearance of infinities ('singularities') in the solutions to his equations. In 1935, in a paper written with Nathan Rosen he states that

> 'A singularity brings so much arbitrariness into the theory . . . that it actually nullifies its laws . . . Every field theory, in our opinion, must therefore adhere to the fundamental principle that singularities of the field are to be excluded.'[13]

His close friend and collaborator Peter Bergmann wrote that

> 'It seems that Einstein always was of the opinion that singularities in classical field theory [i.e. physics] are intolerable . . . because a singular region represents a breakdown of the postulated laws of nature. I think that one can turn this around and say that a theory that involves singularities and involves them unavoidably, moreover, carries within itself the seeds of its own destruction.'[14]

What these authors are so worried about is the fact that if a physical infinity appears in a theory like Einstein's, where the fabric of space and time is determined by the physical density of matter within it, then it requires that space and time are destroyed at the places where infinite densities appear. This means that the laws of gravity will cease to hold at physical infinities and the goal of science to predict the future will become impossible there. This is why physical infinities are a so much more serious business than the problem of mathematical infinity, which is just a matter of policy.

Not everyone follows this desire to avoid infinities at the beginning of the Universe at all costs. If you are looking to the infinity at the beginning of the Universe to be the 'hand of God', then you don't mind the laws of physics being broken, suspended or transcended there. This is just where God lights the blue touchpaper (and retires[15] according to the Deistic picture). However, like Hawking, those who seek to exorcise the infinities implicit in the traditional big bang beginning to the Universe do not necessarily want to overthrow the idea that the Universe had a beginning; they merely seek to have a beginning that occurs with the Universe having finite qualities which are amenable to description by laws of Nature.

There are others who see the initial infinity as an essential part of the physical description of the Universe. Roger Penrose argues that the infinities of Einstein's theory that seem to mark the beginning of the Universe[16] are fundamental and will not be removed[17] by a deeper

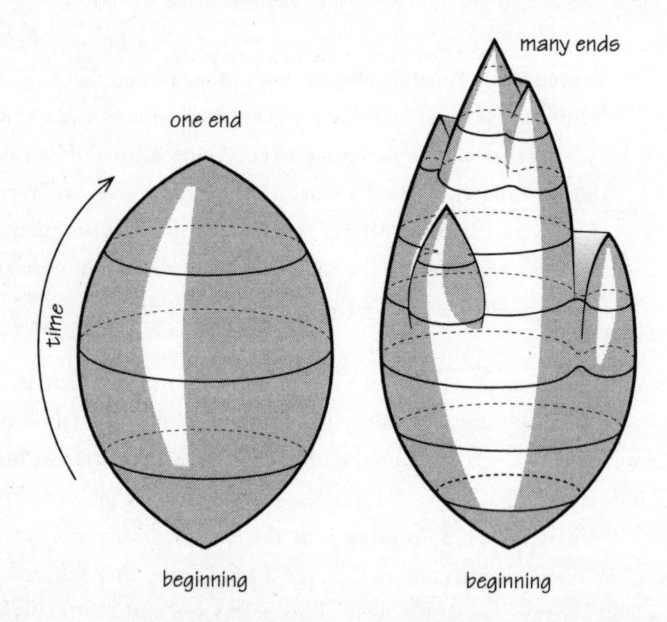

Fig 6.5 *A perfectly ordered universe has a simultaneous beginning and a simultaneous end. A more realistic universe will have irregularities that grow at different rates in different places and so different parts of the universe will experience many local ends of infinite density at different times.*

theory, although their infinite character might be changed in some fundamental way. He believes that infinities at the beginning and the end of the Universe's history (see Figure 6.5) have quite different structures which are an essential reflection of the inevitable evolution from order to disorder that we call the 'second law of thermodynamics'.

NAKED INFINITIES

> 'angelheaded hipsters burning for the ancient heavenly connection to the starry dynamo in the machinery of the night.'
>
> Allen Ginsberg[18]

The beginning of the Universe – if indeed it had a beginning – is a unique moment. What happened then is likely to be controlled by principles that need have no application at any other place and time in cosmic history. We could regard the beginning of the Universe as an event of such a special sort that we are going to exclude it from the case for and against physical infinities. Instead, we should ask if there can be physical infinities in the Universe today: infinities that we could see.

This is a question that Roger Penrose played a key role in formulating precisely and pointing us towards possible answers. If a cloud of material is larger than about three times the mass of the Sun, then it can keep on contracting in size under the force of its own gravity. No known force of Nature can resist its gravity. At first sight, this appears to be a recipe for creating a physical state of infinite density in a finite time almost anywhere in the Universe. However, the situation turns out to be unexpectedly subtle. Once a cloud of the required mass contracts to a particular critical size, it becomes invisible to outside observers. The strength of its gravity is sufficient to stop light passing out beyond a critical surface, or 'horizon', and its interior can no longer be seen. The outside astronomer can feel the pull of its gravity, but know nothing of events inside the horizon. This situation describes the formation of a 'black hole'. From the outside observer's point of view, the black hole looks like an unchanging gravitational field.[19] A large black hole looks red rather than black because the light that reaches a distant astronomer from just outside the horizon surface will have lost so much energy climbing out of the very strong gravitational field of the black hole that it will have been reddened in colour.

Despite the folklore, it is important to appreciate that black holes are not necessarily solid objects. Large ones, of the sort that seem to lurk at the centre of galaxies, will be nearly a billion times more massive than our Sun, but their density will be less than that of air. We could pass through their horizons right now and notice nothing odd or extreme about our local conditions. Only if we tried to reverse our path and return to base far away would we find it impossible to pass out through the horizon surface (see Figure 6.6).

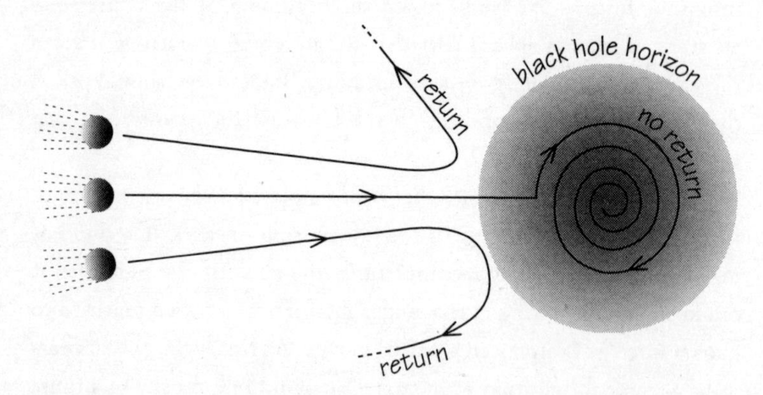

Fig 6.6 *Once an astronaut passes through the event horizon surface of a black hole the journey becomes irreversible. The spaceship cannot return through the horizon surface and no signals can be sent back through the horizon to listeners back home.*

Although the outside observer sees nothing of what lies inside the horizon of the black hole, there is plenty going on there. We could be inside a gigantic black hole right now and notice nothing unusual for a long time. But, gradually, as material keeps falling in towards the centre of the black hole, the local density will keep on rising. Eventually, it will either hit a 'singularity' where the density is physically infinite, or new physics will arise, just as in the discussion of singularity at the apparent origin of the universe, and halt the compression at a very high, but finite, density.[20]

Let us for a moment suppose that a real physical infinity, where the laws of Nature break down, does form at the centre of the black hole. This would be visible to anyone inside the black hole falling in towards the central singularity. They would experience the effects of a physical infinity and there would be no way of predicting what might emerge from the singularity.

Things are completely different for outsiders, though. If you are outside the horizon surface, then you cannot see the central singularity and its effects cannot influence you and me. The physical infinity is cloaked by the horizon and cannot alter the predictability of the laws of Nature in the outside world. Physical infinities would not be visible outside the horizon.

This appealing state of affairs led Roger Penrose to propose that here exists a principle of 'cosmic censorship' in Nature, so that all singularities, or physical infinities, where the laws of Nature break down, are hidden from the outside universe by horizon surfaces. Their consequences are trammelled up by the extreme curvature to space and time that accompany the formation of very high density regions. They are quarantined by the horizon.

There have been many attempts to prove that this hypothesis of cosmic censorship is always true: that naked singularities never occur in Nature, they are all hidden by horizons. So far, it has not been possible to prove that it is universally true, but all the plausible situations that appear to threaten it have turned out to fail. It continues to be suspected that it will turn out to be true, but with certain caveats.

The first type of caveat is to eliminate a number of weird situations that would never arise in the real world because they would require the cosmic equivalent of a needle balancing on its point or a ball rolling up hill with exactly the right speed so as to come to rest on the top rather than running over the top or back down the slope.

Unfortunately, although these pathological situations don't happen in Nature – they require a perfectly tuned physical situation to arise which is extraordinarily unlikely to arise naturally – they tend to turn up in the studies that mathematicians do of these situations. This is because the very special situations they describe are the easiest solutions of Einstein's complicated equations for us to find. So, even though we can find solutions of Einstein's equations which display time travel or the formation of a naked physical infinity, that is not enough to convince us that they exist. We need to know if these special solutions are physically realistic and whether they are stable – if you change them very slightly, do they cease to display the feature you are interested in? We will not observe in Nature sequences of events which are unstable even though they are possible in principle and do not violate any law of Nature. One example would be the sudden coming together of fragments of glass to produce a wine glass – the time reverse of the process of breaking a glass into pieces.

The other caveats all hinge upon the role that might be played by quantum theory. Up until 1974, black holes were believed to be inescapable matter traps. Once you passed in through the horizon there was no escape. Then Stephen Hawking predicted that black holes should not be completely black. Their strong gravitational fields will gradually produce pairs of particles close to the horizon at the expense of the mass and energy of the black hole. Gradually the mass of the hole will evaporate away. The process is very slow for large black holes that exist in the Universe today, and has no effect that we can see. However, if very small black holes, about the mass of a large mountain and the diameter of a single proton, had formed billions of years ago, they would be in the final explosive stages of their evaporation today. We would see them exploding in the Universe today, giving out their energy in radiation and fast-moving particles.

The black-hole evaporation process introduces a new type of physical process that was not included in the original spectrum of possibilities that would allow cosmic censorship to hold sway. The horizon will steadily shrink to zero size as the mass of the black hole completely disperses. But what remains afterwards? Cosmologists don't know. Some argue that 'nothing' remains. Others suggest that a local physical infinity remains, like that at the beginning of the Universe. And there are others still who suggest that a stable relic mass remains[21] and the evaporation cannot continue whittling away the black hole's mass all the way to zero.

If the physical infinity really did form, then it would be visible to outside observers and its unpredictable effects could influence us. This would show that when quantum physics was added to the game, then cosmic censorship does not hold. Yet this conclusion is not widely accepted. Just as the addition of quantum theory to our description of events in the early Universe is expected to smooth out physical infinities into events of very high but finite density, so the same is expected in the high-density remnants of black hole explosions, if they occur.[22]

What we have found is that, despite the special attitude of cosmologists to the existence of physical infinities, and the particular places where they would expect to find them, there is a general unwillingness to admit them into the Universe because of their unpredictable consequences. Rather it is expected that their prediction is another signal that existing theories need more work in order to extend their domain of applicability. Physical infinities will be nuanced by the laws of Nature.

THE GREAT BLUE YONDER

'It is a great advantage for a system of philosophy to be substantially true.'

George Santayana[23]

The third flavour of infinity is the most familiar, the most controversial, and the least amenable to investigation. To some it is a matter of faith, to others a state of mind, and to most others a harmless mystical feeling about the Universe that does not have any real impact on the here and now. This is what we might call transcendental or, using Cantor's words, absolute infinity. It is the cosmic encompassment of everything. For some it is a necessary attribute of God. A typical statement about their close association is that 'God is infinite in His nature, unlimited and unbounded in every positive way.'[24]

We see immediately why the history of mathematical inquiry into the nature of the infinite has been so fraught and dangerous for those indulging in it. It is like the creation of a false god, or an attempt to describe God in a limiting way, or to deny God the property of uniqueness. Cantor's demonstration that there was no largest infinity appealed to some theologians of his day precisely because it left open a place for a God who was greater than any quantity that could be named or defined. It is not clear whether such a characterisation of God has much useful overlap with the picture provided by human monotheistic traditions. It is easy to imagine what being infinite in time might mean – eternal, always existing to the past and the future – because we are faced with theories of the astronomical Universe which require it to be infinitely old as well. It is harder to imagine what it means for the Deity to be infinite in extent. Instead, theologians are more interested in the lack of bounds on particular attributes; the absence of limitations of different sorts; or simply being greater than any human imagining:

'Nor can one speak of [God] as having parts, for that which is "One" is indivisible and therefore also infinite – infinite not in the sense of measureless extension but in the sense of being without dimensions or boundaries, and therefore without shape or name.'[25]

In the Russian Orthodox tradition of apophatic theology, the infinite nature of God is a natural consequence of the inability in its negative theology to say what God is, only what He is not.[26] This contemplation of absolute infinities has also indirectly led to a particular subtle type of argument for the existence of God that is more than a thousand years old.

The collection of arguments for the existence of God that are known as the 'ontological arguments' began in 1078 with the argument of Anselm (1033–1109), an early Archbishop of Canterbury. Anselm implicitly used the infinity of God to endow God with certain properties which he claimed made His existence logically necessary. He argued that if God is the greatest conceivable being, then He must have *actual* existence, not just potential existence, for otherwise a more perfect being could be conceived of: one that had the added attribute of actual existence.

Arguments like this have a long and chequered history. They are a sleight of hand in that they appear to prove something very strong, but merely show that their striking conclusion is equivalent to a less eye-catching but equally strong initial assumption. They are really just showing that if it is possible for a perfect and omniscient being to exist, then such a being necessarily exists. But that first 'if' is a pretty big one and demands at least as much proof as the existence of God.

These existence 'proofs' can also have unexpected consequences. If one defines 'God' as having all qualities, and existence is a quality, then God must have the quality of existence and so exist. But, equally, non-existence is a quality and God must have this quality too, and so must not exist. As Immanuel Kant first noticed, the problem with this type of argument is that existence is assumed to be a property, whereas it is really just a precondition for something to have properties. 'Some dogs are black' makes sense because colour is a property of dogs, but 'some dogs exist' does not make sense because existence isn't a property of dogs.[27]

Here is another argument of this type where the infinite aspect is clear. It is called the argument from omniscience, or infinite knowledge.[28]

Assume that God knows everything. He know all things that are true and all things that are false. Assume also that God is rational. Hence, He believes in His own existence whether or not He actually exists, just as Sherlock Holmes believes in his own existence even though he does not exist. If He does exist, He is correct in this belief in His own existence; if He doesn't exist, then he is mistaken. But if God did not exist, His assumed omniscience would require Him to know that

---

Kurt Gödel's Ontological 'Proof' of the Existence of God

Axiom 1: A property is positive if and only if its negation is negative.

Axiom 2: A property is positive if it necessarily contains a positive property.

Theorem 1: A positive property is logically consistent (that is, possibly it has an existence).

Definition: Something is God-like if and only if it possesses all positive properties.

Axiom 3: Being God-like is a positive property.

Axiom 4: Being a positive property is logical and hence necessary

Definition: A property P is the essence of x if and only if x has the property P and P is necessarily minimal.

Theorem 2: If x is God-like, then being God-like is the essence of x

Definition: x necessarily exists if it has an essential property,

Axiom 5: Being necessarily existent is God-like.

Theorem 3: Necessarily there is some x such that x is God-like.

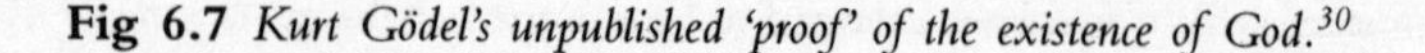

Fig 6.7 *Kurt Gödel's unpublished 'proof' of the existence of God.*[30]

fact. But, that would contradict the assumption of His rationality. Therefore God must exist!

Again, the argument derives the existence of God from an initial assumption that it hopes you will not notice is equally demanding of proof – that a perfect all-knowing being can exist.[29]

Even the great logician Kurt Gödel seems to have fallen into this trap with his unpublished version of the ontological argument, see Figure 6.7.

INFINITY ON THE BACK FOOT

'If you stare too long into the abyss, the abyss will stare back into you.'

Friedrich Nietzsche[31]

Our tour through the present uses and abuses of infinity has found that Cantor's mathematical infinity has become an uncontroversial part of modern mathematics, 'a paradise', as David Hilbert described it, 'from which no one will expel us'. Yet the realities of physical and absolute infinities, which had a relatively trouble-free course through the philosophy of the nineteenth and twentieth centuries, have lately encountered sterner and sterner opposition from physicists and philosophers. Infinity has increasingly become the touchstone for the failure of physical reasoning, the signal that the mathematical theory you have been using to do physics is out of its depth. It needs attention. This creates an awkward dilemma for scientific inquiry into the frontiers of physical reality. Unless we know that physical infinities cannot occur, then using their appearance as a monitor for theory breakdown will fail us in some circumstances. We need a finer grading of the types of infinity that occur in physical descriptions of the world, just as we have in the catalogue of mathematical infinities.

chapter seven

Is the Universe Infinite?

'Our minds are finite, and yet even in the circumstances of finitude we are surrounded by possibilities that are infinite, and the purpose of life is to grasp as much as we can of that infinitude.'

Alfred North Whitehead[1]

EVERYTHING THAT IS

'I overthink things. I should have been a Greek philosopher, but I didn't have the brains.'

Badly Drawn Boy[2]

One of the questions that first drew people to think about the notion of the infinite was the problem of the Universe: everything that is. Does the Universe just go on forever or does it have an edge? The same questions were asked about the Earth as well, and there were cultures who still believed that the Earth was flat at the same time that others had deduced that it was curved (see Figure 7.1). Even when the special features of a near-spherical Earth were common knowledge – allowing you to sail forever without falling off the edge of the Earth – the same ideas were not carried over into our picture of the Universe. There was no obvious way in which the Universe of space could be thought of as

Fig 7.1 *Some imaginary flat and curved 'Earths'.*

the curved surface of a ball. But, we shall see, our current understanding of the Universe provides us with other more unusual possibilities.

In most old cultures there were systems of belief which incorporated a theory or a legend about the nature of the Universe and our position within it. These beliefs, perhaps in the form of creation myths or stories about how the world remains in being, had an important psychological role to play. They gave humanity a meaningful place within the cosmic scheme of things. They pushed back the boundaries of the unknown to places where they could have no immediate impact upon what happened here and now. In this context the question of whether the Universe went on forever, or whether it came to a stop, was one that needed an answer that fitted in with beliefs about other things.[3]

The first of the modern European astronomers to pursue the idea that the Universe might be infinite in size was the English astronomer Thomas Digges (1546–95). Digges, a scientist and military scholar, was one of the few early supporters of Copernicus's new

heliocentric model for the solar system. In 1576 he published a book, *A perfit description of the caelestiall orbes,*[4] which used Copernicus's system for the motions of the Sun and planets and also proposed that the Universe is infinite in extent. He was the first astronomer to take that step. Before him, cosmological models of the sky lay inside a spherical shell of stars. Beyond the shell's outer edge lay 'Paradise' and the domain of the 'Prime Mover'. Digges did away with that outer boundary, replacing it by a space of unlimited extent that was filled with stars. Digges was the first of the Renaissance scientists to propose that the Universe itself was physically infinite. He was careful to use the Universe's infinite nature to reflect the greatness of God, so that

> 'we may easily consider what little portion of God's frame our elementary corruptible world is, but never sufficiently be able to admire the immensity of the rest, especially of that fixed orb garnished with lights innumerable and reaching up in spherical altitude without end.'

His universe is shown in Figure 7.2, a famous image that formed the frontispiece of his book. It bore the grandiloquent caption:

> 'This orb of stars fixed infinitely up extendeth itself in altitude spherically and therefore immovable, the palace of felicity garnished with perpetual shining glorious lights innumerable far excelling our sun both in quantity and quality, the very court of celestial angels devoid of grief and replenished with perfect endless joy, the [home] for the elect.'

The Sun is at the centre, encircled by the six planets (the large circular badge three from the centre is the Earth, with Mars, Jupiter, and Saturn beyond). The outermost known planet at that time was Saturn.[5] We

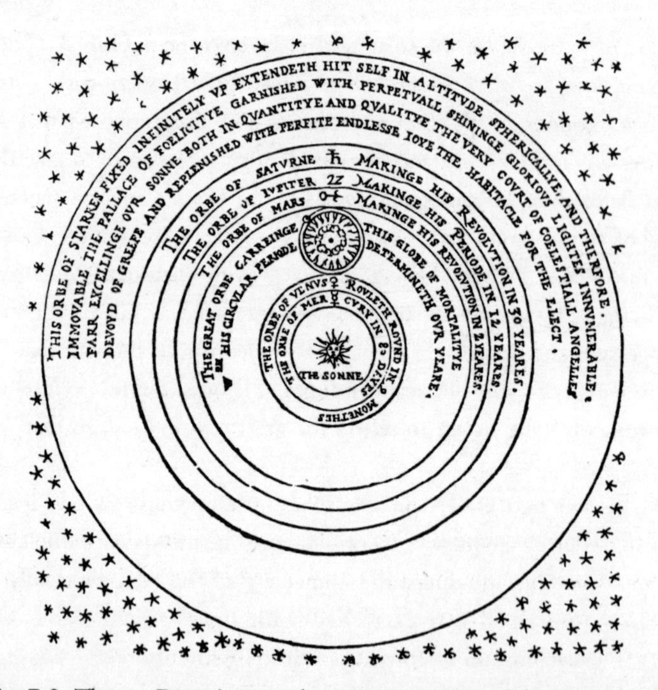

Fig 7.2 *Thomas Digges' sixteenth-century conception of the Universe.*

see that beyond there is a large empty region before we reach the realm of the fixed stars which are imagined to go on forever.

Digges was a contemporary of William Shakespeare (1564–1616) whose writing career intersected a time of great intellectual ferment – the Renaissance, the Reformation and the gradual confirmation by Galileo of Copernicus's model of the Universe. In fact Shakespeare knew the Digges family and was acquainted with the correspondence between the great Danish astronomer Tycho Brahe and Digges's associates, and the portrait of Tycho under the family shields of his great-great-grandparents Sophie Gyldenstierne and Erik Rosenkrantz. Tycho still supported an Earth-centred model of the Universe which he published in 1588 and it appears that Shakespeare constructed *Hamlet,* his greatest tragedy, with all manner of allusions to these astronomical debates and the protagonists involved.[6]

His characters Rosenkrantz and Guildenstern emerge in this way, representing the geocentric world-view of Tycho, which the false king Claudius derives from the ancient astronomer Claudius Ptolemy (c. AD 140), whose model of the Universe was displaced by Copernicus. Claudius memorably summons the two courtiers to help him with a new geocentric model, whereupon Hamlet makes his eloquent poetic affirmation of the infinite world of his friend Thomas Digges: 'I could be bounded in a nutshell, and count myself a king of infinite space.'[7]

The continental contemporary of Digges was the infamous Giordano Bruno (1548–1600). He is remembered primarily for dying as a martyr to his belief in the infinite Universe, but he was in no sense a scientist. Born and raised in Nola in Italy, he entered the Dominican monastery in Naples as a teenager and there became familiar with the new astronomy of Copernicus. Eventually he turned into a full-time iconoclast and itinerant philosopher who preached his heretical views all over Europe, making enemies within the Catholic hierarchy (and in many other places as well) wherever he went. For him, Copernicus was a symbol of opposition to the powers that be and the inflexible traditions that shackled them. Bruno dreamed of expanding

Copernicus's system yet further. He wanted a whole infinite Universe full of stars like our Sun, each surrounded by its own system of planets on which intelligent beings could live:

> 'Thus is the excellence of God magnified and the greatness of his kingdom made manifest; he is glorified not in one, but in countless suns; not in a single earth, but in a thousand, I say, in an infinity of worlds.'[8]

His watchword was infinity, but his writings are a strange mixture of the mystical, the confused and the insightful. Time and space were both infinite in his conception of things, and he saw that an infinite body can have neither centre nor boundary. Thus amongst his never-ending collection of worlds the Earth 'no more than any other is at the centre'.

In 1591 Bruno was an unsuccessful candidate for the professorship of mathematics at the University of Padua that was awarded to Galileo the following year. After that he seemed to face increasing criticism and persecution from the authorities for his strongly anti-Aristotelian views. Here is part of his dialogue between Philotheo, Fracastoro and Elpino that pokes fun at Aristotle's belief in the necessity of a finite Universe:

> '*Philotheo*: If the world is finite, and if there is nothing beyond the world, then I ask you: *Where* is the world? *Where* is the universe? Aristotle's reply is: The world is in itself. The convex surface of the primordial heaven is universal space, and as the primordial container is not contained by anything else; for location is merely the containing body's surfaces and limit, so that he who has no containing body has no location. But, dear Aristotle, what do you mean by the "location is in itself"? What will you tell us about that which is beyond the world? If you say there is nothing, then the heavens and the world will surely not be anywhere at all.

> *Fracastoro*: Therefore the world will be nowhere. Everything will be in nothing.'

These lines would not have looked out of place in Shakespeare's *Much Ado About Nothing*. Bruno saw clearly that an infinite Universe needs no centre and Copernicus's philosophy requires no special places – no bounding spheres or special places – just one Universe, everywhere looking the same, without boundary, and full of an infinite number of stars and planets. His spokesman lays out his philosophy in a nutshell:

> '*Philotheo*: All things then are one: the heavens, the immensity of space, our mother earth, the encompassing universe, the ethereal region through which all things move and continue on their way. Herein our sense may perceive innumerable heavenly bodies, stars, spheres, suns, and earths; and reason may deduce an infinitude of them. The universe immense and infinite, is the sum total of all that space and all the bodies it contains.
>
> *Elpino*: So there are no orbs with surfaces concave or convex, no deferent circles. Instead, all is one field, a single common envelope.
>
> *Philotheo*: That is right.'[9]

Bruno foolishly accepted an invitation to come to Venice to act as the tutor to Giovanni Mocenigo, an agent of the Roman Inquisition who claimed to be seeking instruction from Bruno in astronomy and in the art of memorisation – a special skill of Bruno's. He seems to have taught his astronomical views a little too clearly to his supposed pupil and, not surprisingly, he was arrested and tried for heresy. He was burned alive at the stake in Venice on 17 February 1600 and never recanted his beliefs.

COSMOLOGY GOES UNDERGROUND

'Oh dear, what can the matter be?'

Folk song[10]

Astronomers are still interested in the question of whether the Universe is finite or infinite, but they recognise that there is a raft of subtleties attached to this question. In 1915, Albert Einstein provided us with a new theory of gravity that could describe the Universe as a whole. This theory introduced a new conception of space and time. Both are fashioned by the distribution and motion of the mass and energy they contain. Universes that contain too great a density of matter will have their geometries curved up into a finite volume while emptier spaces can extend unfurled, forever. Their continuation in time may be limited too. At a finite time in the future their expansion should slow down and gradually be replaced by contraction towards some future Big Crunch of enormous density and temperature. In sharp contrast, those universes that possess a density not exceeding a certain critical value will be able to expand forever (Figure 7.3), becoming sparser and more rarefied.

The density that marks this critical divide is quite low by terrestrial standards – only six atoms in every cubic metre of space. This is far emptier than any 'vacuum' that we can produce artificially in laboratories on Earth. If we take all the matter that we can see in the Universe – shining in optical light or emitting other forms of radiation, like X-rays – then we have so far discovered a total that amounts to a density of just one atom in every seven cubic metres. This is well short of the total needed to make up the critical density. Yet, we can't just jump to the conclusion that the Universe is infinite. Lots of matter might be invisible to our telescopes and detectors because it is cold and dark.

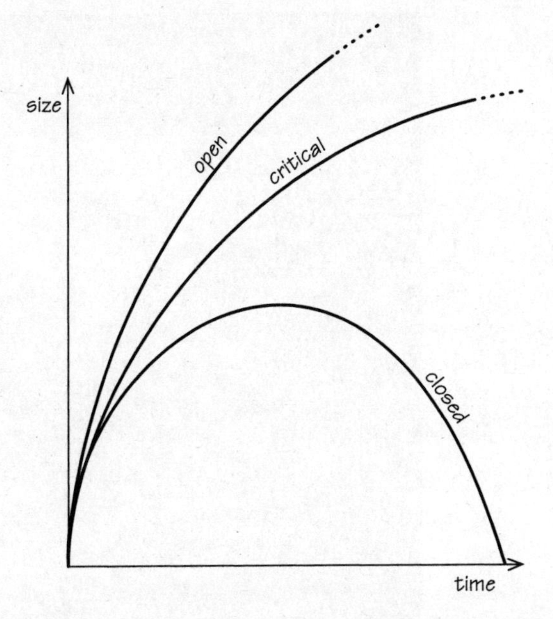

Fig 7.3 *Some expanding universes expand forever whilst others will ultimately contract. The 'critical' divide is characterised by the special universe that is just able to keep expanding forever. If it contained any more matter or expanded more slowly then it would eventually contract.*

Look at the Earth from space, at places where it is not illuminated by the Sun, and the appearance of light there will not necessarily tell you where all the people are. It will tend to tell you where all the money is. The big cities, like London, New York and Tokyo, will emit lots of light (see Figure 7.4). But the densest population centres in Africa and China will be almost entirely dark.

The lesson: light is not necessarily a reliable tracer of population density – and so it seems to be with the population density of matter in our Universe. Light is emitted from the places where the density of matter has grown largest. These are the places with more matter than the average; they exert greater gravitational pull on neighbouring matter and become even denser at the expense of the rarefied regions. They

are the crests of the waves in the universal sea of matter. They are the densest and most visible places in the Universe.

But what's in between them? Fortunately, we have ways of probing the dark voids between the shining stars. Everything, whether it shines or not, should have a gravitational effect on other matter. By watching how fast the luminous stars and galaxies are moving, we can determine the strength of the gravitational forces they are feeling. Remarkably, wherever we do this we discover that things are moving as if they are under the influence of the gravity of about ten times more matter than we see shining in the dark. We call this other unseen material 'cold dark matter'. A small fraction of this dark matter is composed of ordinary atoms and molecules, but the identity of the rest is a mystery (see

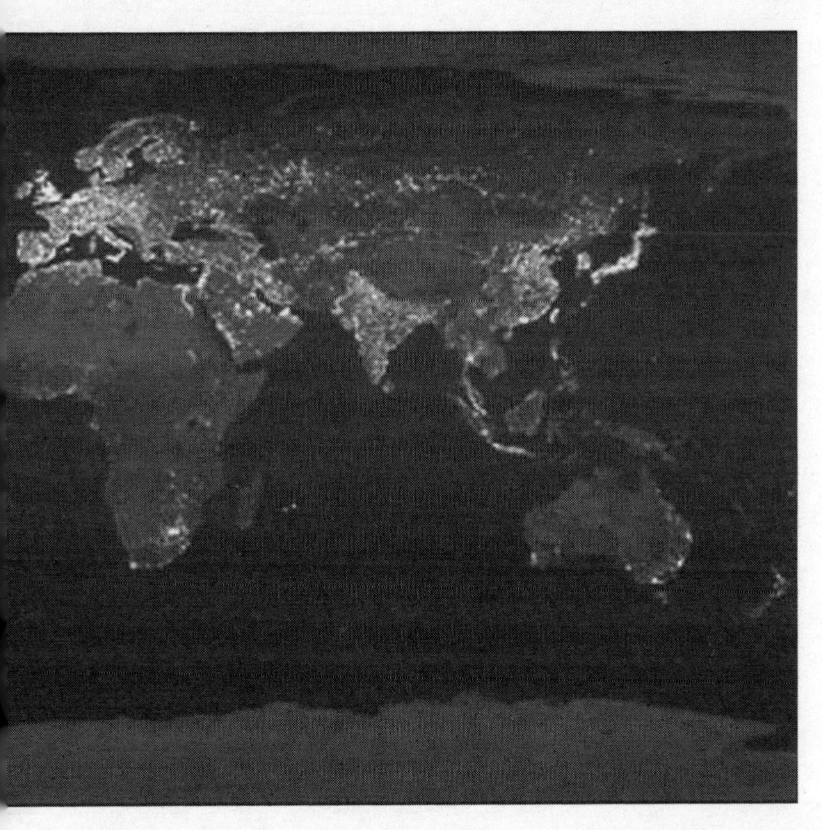

Fig 7.4 *A composite satellite photograph of the Earth at night.*[11] *Notice that the most brightly illuminated areas are big Western cities. The regions of greatest population in Asia and Africa are almost completely dark. Thus the light is a good tracer of wealth rather than of people.*

Figure 7.5). Two of the great quests of modern cosmology are to pin down the quantity and the quality of the cold dark matter.

Our first guess about all this dark material pushing and pulling our stars around is that it's the same stuff that goes into planets, stars and galaxies. It just hasn't found itself in clumps dense enough to begin contracting, initiate nuclear reactions, and start shining as 'stars'.

Alas, things are not so simple. If we try to explain away the dark

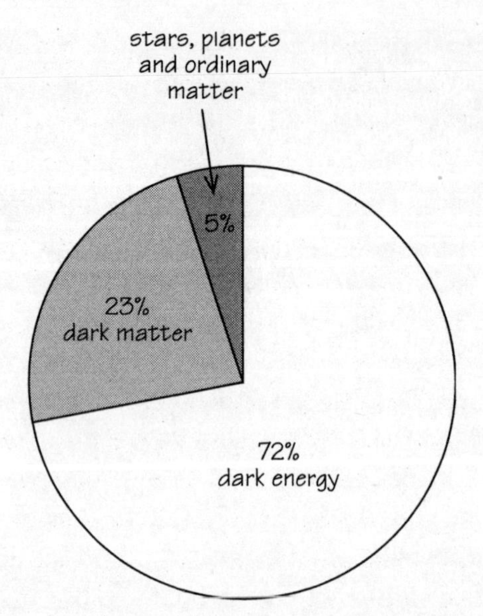

Fig 7.5 *The material composition of the universe showing the fractions in the form of luminous matter, dark matter, and an unidentified form labeled 'dark energy' which is responsible for accelerating the expansion of the universe in the last few billion years. The dark energy displays gravitational repulsion whereas all the other forms of matter and radiation we detect display gravitational attraction.*

matter in this way, we run into a serious conflict with some of our other evidence. Ordinary matter is made of atoms which contain nuclei of protons and neutrons. These particles can take part in strong nuclear reactions and build up heavier nuclei step by step from the simplest – which is the hydrogen nucleus comprising a single proton – to make nuclei like deuterium (one proton + one neutron) or helium (two protons + two neutrons), and beyond.

When the Universe was just two minutes old, it should have been hot enough for nuclear reactions to occur everywhere, transforming the protons and neutrons produced earlier[12] into definite abundances of deuterium, helium and lithium. These abundances are quite straightforward to calculate. Even though a universe that is only two minutes old sounds bizarre, the properties we expect it to have then are not outlandish. The density of matter will be little more than that of water. We are not talking about conditions so extreme that we lack a good understanding of the laws of physics that govern them.

Remarkably, not only can we calculate the abundances of the lightest elements that emerge from the early stages of the Universe, but those abundances agree with those that we measure in our own Galaxy and in others. The abundances of deuterium, the two isotopes of helium – helium-3 and helium-4 – and lithium are beautifully explained by the inevitable sequence of nuclear reactions that took place in the first few minutes of the Universe's expansion. Last, but not least, we know of no way of making these particular elements in their observed abundances by other astronomical processes in the stars, whereas such processes are able to explain the abundances of all the other heavier elements in the Universe.

What has this got to do with dark matter? Well, we get good agreement between the observations of the lightest elements in the Universe and the predictions of their production in the early Universe if we use the known abundance of ordinary matter to determine the nuclear reaction rates when the Universe was two minutes old. If we try to pass the mysterious dark matter off as yet more ordinary matter, then we run into a big problem. Increasing the number of ordinary protons in the Universe in this way speeds up the nuclear reactions during the first three minutes. The abundance of helium-4 is not greatly changed, but the amount of deuterium and helium-3 is significantly reduced below what is seen.[13]

Our observations are therefore telling us that the bulk of the matter in the Universe is in some unknown form that cannot participate in nuclear reactions. It is not made out of atoms and molecules as we are, so what could the dark matter be?

There are many candidates which fit the bill. They need to be abundant and they must take a form that does not participate in nuclear reactions. Elementary particles like neutrinos are ideal candidates. They feel only gravity and the weak force that is responsible for radioactivity and they are expected to emerge from the early stages of the Universe in interesting quantities. At first, in the early 1980s, cosmologists thought that the lightest known neutrinos would be the prime suspects. There was growing evidence that they possessed a tiny mass, about ten billionths of the mass of a hydrogen atom, and that seemed to give just the right density of neutrinos to account for the mysterious abundance of cold dark matter today.

Unfortunately, over the last twenty years it has become clear that the known types of neutrino, of which there are three, cannot be the cosmic dark matter. A combination of laboratory experiments and observations of stars where neutrinos are produced has constrained their masses to be so low that they are almost certainly not the cosmic dark matter. Equally problematic have been the studies, using some of the most powerful computers in the world, of what would happen to the luminous matter if it felt the gravitational pull of a universe of lightweight neutrinos. It becomes impossible to get the distribution of matter to form clumps and galaxies on much smaller scales. Since the three neutrino species that we know about can only have very small masses, they move very fast in the Universe. In order to cluster into galaxies and pull matter into clumps where stars form and light shines, they need to move much more ponderously. We need neutrinos that are heavier and slower.

So what is left? The strange dark matter could be composed of very small black holes, each no more massive than the Earth but only one centimetre in size. Black holes don't take part in nuclear reactions and so don't upset our predictions about the lightest elements. They could perhaps be the dark matter – but it seems odd that they should form so abundantly with the mass of the Earth. Why this mass? If more massive black holes were equally abundant it would contradict other observational restrictions. So, there is a credibility gap. Why should black holes happen to have formed with just the right masses and abundances to solve our dark matter problem? In the absence of a good reason cosmologists have noted this possibility but not pursued it enthusiastically.

The most popular alternative is that there are other neutrino-like particles in Nature which are much heavier than the known neutrinos. Like the known neutrinos they will only take part in weak interactions with other elementary particles of matter and like all forms of matter they will feel the force of gravity. Although such particles have not been detected directly, they are expected to exist. Their masses could lie anywhere between the mass of a proton up to many thousands of times its mass. A definite calculation of the masses of these particles has not yet been possible. We just have to look carefully to find them.

In several sites around the world there are large underground detectors dedicated to finding these heavy particles. If they do constitute the dark matter in and around our Galaxy then they will frequently be flying right through the Earth and emerging from the other side. Although their interactions with ordinary matter are very weak, they will occasionally rebound from the nucleus of an atom of silicon or xenon, shaking it up and leaving it a little more energetic than before. Ultimately, we hope to be able to detect the tell-tale recoils produced by the heavy neutrino-like particles as they pass through the Earth. If they are abundant enough to explain the dark matter's gravity, and they move with the speed that gravity's pull requires, then they should hit our detectors often enough each day to be found in the next few years.

If we follow the history of these heavy neutrino-like particles with our most powerful computers, then they are much more interesting than lightweight neutrinos. Their heaviness means they move more slowly and they can condense into regions small enough to explain the existence of cold dark matter in galaxies. This is an exciting frontier of modern cosmological research that brings together particle physicists with their candidates for the cold dark matter particles, astronomers with their observations of how much dark matter there seems to be, computational astrophysicists running huge computer codes to simulate the formation of galaxies dominated by slow-moving dark matter, and experimental physicists searching for the tell-tale signatures of the dark matter particles flying through their detectors deep underground.

Fig 7.6 *Maurits Escher's beautiful woodcut of a loxodrome,* Sphere Spirals *(1958). The curve cuts each of the lines of longitude at the same angle and this results in an infinite spiralling into the North and South Poles of the globe.*[14]

Until just a few years ago the observations pointed stubbornly to a Universe that did not contain the critical density of material needed to halt its expansion in the future. Even the dark matter couldn't tip the scales. On the simplest interpretation of the evidence it appeared that the Universe could not be finite. Or could it?

BENT UNIVERSES

'I used to measure the heavens
Now the earth's shadows I measure
My mind was in the heavens
Now the shadow of my body rests here.'

Epitaph on Kepler's Grave[15]

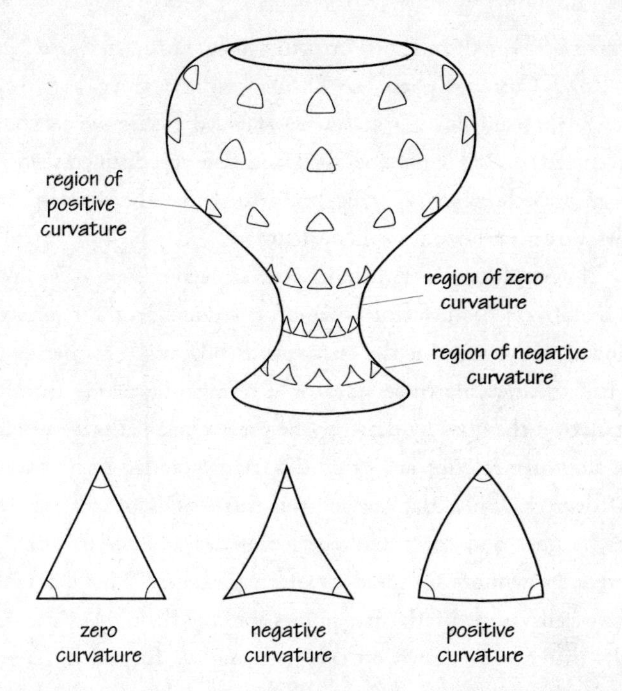

Fig 7.7 *The shortest distances between two points at different places on the surface of an inverted decanter. The bulbous region at the top has positive curvature; the area near the lip has negative curvature; in between there is a place where there is no curvature and the surface is flat.*

We are familiar with surfaces that are curved in some way. The surface of your hand is curved. Einstein taught us that the presence of matter in space curves it in a particular way. If the material moves around, then the curvature of the space will change as well. The presence of curvature determines what is the shortest distance between two points. On a flat surface this is a simple straight line, but on a curved surface the shortest line is less obvious. In Figure 7.6 we see Escher's beautiful woodcut of the loxodrome or rhumb line on the Earth's surface. It cuts each of the lines of longitude at the same angle. Until the middle of the sixteenth century, navigators believed that following the rhumb line joining two points on the Earth's surface gave you the shortest distance between them. The Portuguese mathematician Pedro Nunes,[16] the Royal Cosmographer, was charged with devising a navigational strategy that did not rely upon new-fangled globes which could fall into enemy hands. He invented the loxodrome and distinguished it from the great circle which we now know to define the shortest distance between two points on a spherical surface.

If you pick up a traditional glass decanter (Figure 7.7) you can see the effects of different types of curvature. In the region of the round bowl we say that the curvature is positive. This means that if we had to draw the three sides of a triangle by taking the shortest distance on the glass by drawing the great circle paths we would find that the three interior angles of the triangle added up to more than 180 degrees. On a flat surface the three sides of the triangle are straight lines and the three interior angles add up to exactly 180 degrees. Now move up the neck of the decanter. This is a region of negative curvature. Mark three points and join them up by the shortest paths that can be drawn on the glass and we form a different type of triangle: one whose three interior angles add up to less than 180 degrees.

The curvature of space is a property of its geometry and it is this feature that is determined by the presence and motion of matter in Einstein's theory of gravity. By contrast, Newton's theory specifies space to be an unchanging stage on which the motion of matter takes place. As we saw in the last chapter, whatever happens to matter, it cannot affect the geometry of space in any way in Newton's theory. Even the cataclysmic future Big Crunch cannot change the geometry of space in any way. Everything in space may be destroyed, but space and time go on forever.

The separation between space and its material contents in Newton's conception of Nature meant that the question of whether the Universe is finite or infinite is not a simple one. Space could go on forever and contain either an infinite amount of matter, dispersed throughout its volume without end, or it could contain only a finite

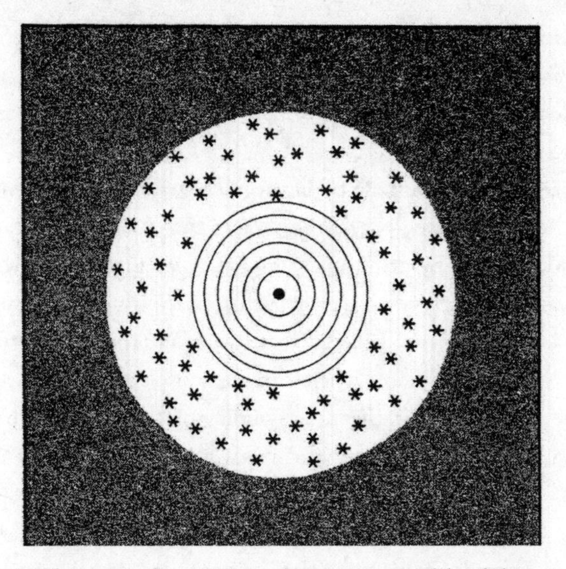

Fig 7.8 *The Aristotelian, Stoic, and Epicurean models of the universe.*

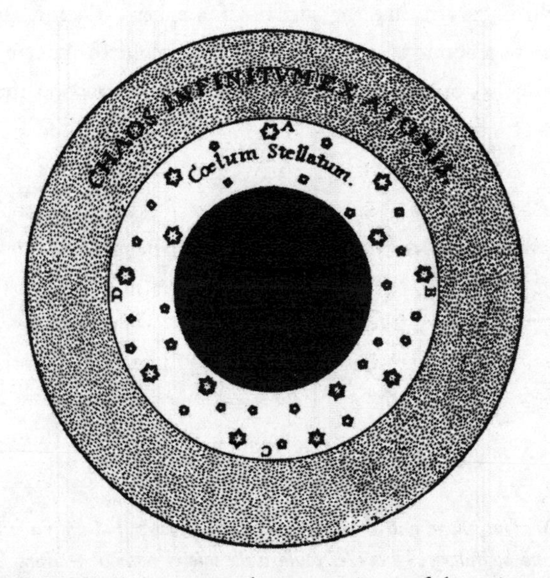

Fig 7.9 *Isaac Newton's seventeenth-century picture of the universe.*[17]

amount. In fact, there are further options. It is possible for the matter to be infinite in extent, but only finite in total mass if matter thins out gradually, leaving space empty beyond a particular distance from the centre (see Figure 7.8).

Indeed, at the beginning of his work on gravity in the seventeenth century, Newton viewed the Universe as a finite system of stars and planets surrounded by an infinite empty space (Figure 7.9). Others, like Descartes, had argued that where there was no matter there could be no space, but Newton believed that the Divine spirit supported the existence of space in places where there was no matter.

The big new idea in the understanding of the difference between finite and infinite spaces came 300 years later, with Einstein's realisation that space could be curved.[18] The most important new consequence is that space can be finite but have no edge. To see how this is possible, consider some examples of two-dimensional worlds. The top of your table is a good example of a flat 2-d world. It is finite and so if a flat world is to be finite it needs an edge. But suppose the 2-d world is curved, like the surface of a sphere. Then it is finite – only a finite amount of paint is needed to colour the sphere – but it has no edge. A Spherelander can keep moving around on the surface

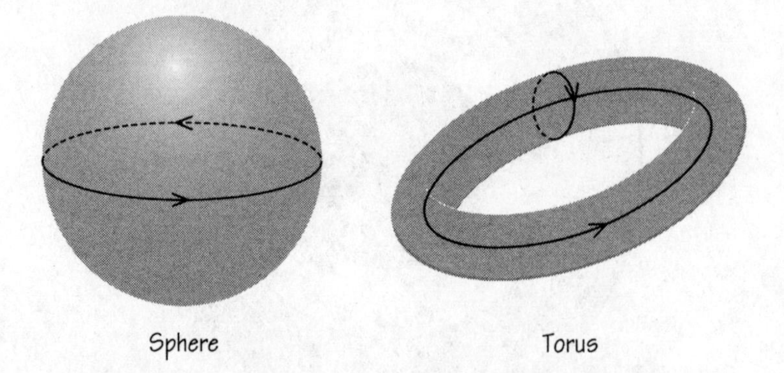

Fig 7.10 *The sphere and the torus both have surfaces that are finite in area but have no boundary. A traveller could walk forever without reaching an edge.*

of the sphere forever without ever hitting a boundary. There are other curved spaces with this property. Take a ring doughnut – what mathematicians call a torus. It also has a 2-d surface that is finite in area and curved in different ways in different places (Figure 7.10).

Curved space therefore resolves an ancient dilemma concerning the alternative to an infinite universe: a space with no edge does not have to be infinite. Just as there are finite two-dimensional surfaces of three-dimensional volumes, so our three-dimensional Universe could be the finite boundaryless curved surface that could form the surface of a finite four-dimensional volume. But it is not necessary for that fourth dimension to exist in physical reality.

Einstein's theory shows how the amount of matter in the Universe determines the curvature of space. Thus, if space is to be curved up into finiteness, we require more than a critical density of matter. This is what the search for dark matter is trying to ascertain. However, in the search we have discovered more possibilities and caveats. Suppose we do not find enough matter in the Universe to make up the critical density. Does this mean that the Universe must therefore be infinite? There are three reasons why the answer has to be 'no'.

THE PROBLEM OF TOPOLOGY

'Instead of being arrested, as we stated, for kicking his wife down a flight of stairs and hurling a lighted kerosene lamp after her, the Rev. James P. Wellman died unmarried four years ago.'

US newspaper apology[19]

There is another property of space that we have to worry about in addition to its geometry. This is called its topology. Unlike geometry,

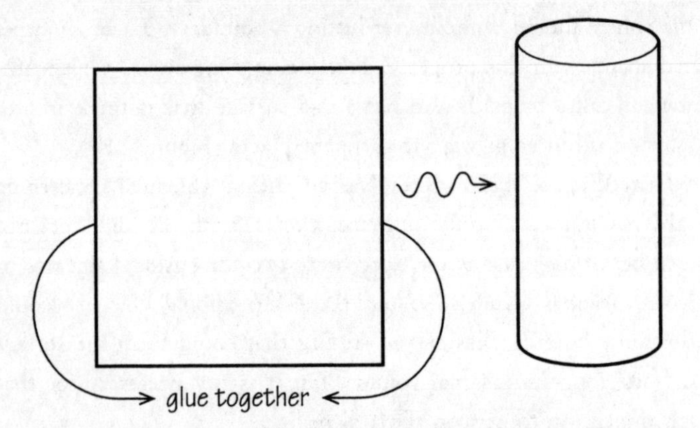

Fig 7.11 *We can wrap up a sheet of paper and glue two edges together to form a cylinder. Its curvature is unchanged. If we cut a hole in the paper we change its topology. The new configuration cannot be turned into the old one simply by stretching and glueing its edges.*

topology is changed only by tearing or cutting holes on a surface or by gluing parts of the space together. So if we pick up a flat sheet of paper and curve it slightly we will not be changing its topology. But if we cut a hole out of the centre or wrap it around to form a cylinder, we will effect a topology change (Figure 7.11).

The case of the cylinder is very interesting. You remember how we detected the curvature of space on the curved surfaces of our decanter by forming the shortest lines between three points. The resulting triangle had angles that added up to more or less than 180 degrees, depending on whether the curvature was positive or negative. And a triangle has angles adding up to exactly 180 degrees if the surface has a flat geometry. Draw a triangle with three equal sides on a flat piece of paper. Now roll the paper into a cylinder and look at the triangle. It is exactly the same! A triangle on a cylindrical surface has three interior angles that add up to exactly 180 degrees. The cylinder therefore has a flat geometry, just like the flat piece of paper on which you drew the triangle. The difference between the flat sheet of paper

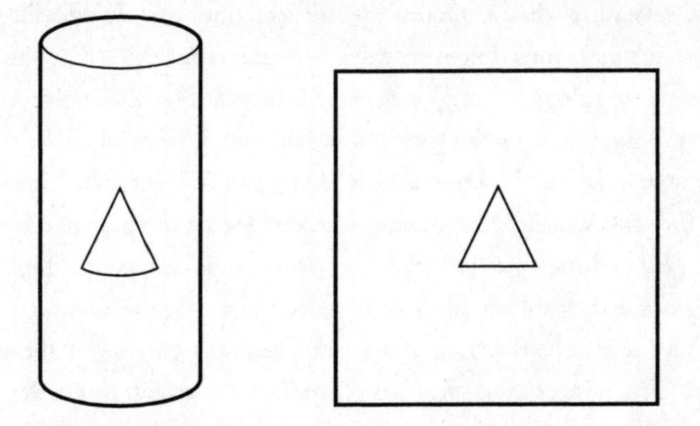

Fig 7.12 *The cylinder is locally a flat geometry. A small triangle created by drawing the shortest distances between three points will have three interior angles that add up to 180 degrees. Although the flat plane and the cylinder have different topology, their geometry and curvature is the same locally.*

and the cylinder is one of topology rather than one of geometry.

This insight makes an enormous difference to cosmology. The simple-minded approach to the expanding Universe assumes that the universes which have spaces with flat or negative geometry have the simplest possible topology – that of a flat or negatively curved sheet that goes on forever. A flat sheet that goes on forever is easy to imagine (at least in two dimensions); a negatively-curved sheet is like a Pringle potato crisp that goes on forever. As a result of this assumption of simplicity, it looks as if negatively-curved universes not only expand forever but they are infinite in all directions.

Nothing could be farther from the truth. Take a geometrically flat universe that expands forever. Roll its space up into a three-dimensional version of the cylinder and the space now has a finite volume. It still has flat geometry everywhere. It still expands forever but it is no longer infinite. Could our Universe be like this?

At present it is very hard to say. Einstein's equations tell us nothing about the topology of space. There must be some deeper law of gravitation that contains the information that is provided by Einstein's equations, but with extra constraints included that show us how the topology of the Universe is selected. The alternative is that the topology is something that just falls out at random. While this is not impossible, it seems a little odd, given the intricate cleverness of the way in which the geometry is wedded to the matter content.

Finite topologies are special in many ways. One might think that this means they are less likely to be selected to describe a universe than infinite ones, but that may not be the best way to look at the question. Their specialness may simply reflect particular properties that single them out as the only ones able to accommodate all the laws of physics within their encompassment.[20] Negatively curved, or 'open', universes allow a vastly greater range of finite topologies with unusual properties. Again, it is not known whether or not we should expect our Universe to be like this. All we can do is go out and see if it is.

If the Universe has one of these interesting finite topologies, then there will be a number of unusual effects on our observations if the diameter of the Universe is too small. We will keep seeing multiple images of the same galaxies over and over again. A universe made finite by its topology is like being surrounded by mirrors. Before the mirrors were in place you received light rays from far away and the longer the light took to arrive so the farther away was the object being seen. Once the mirrors are in place, you see multiple images of any object in between you and the mirror. It looks as if the space around you goes on forever, but you are the victim of an optical illusion. So it is with a finite universe. Multiple images of the same bright galaxies might belie the true size of the universe. More subtly, the statistical distribution of the variations in the intensity of radiation also takes on a distinctive form in topologically finite universes. So far these tell-tale signatures of finiteness without positive curvature of space have not been found. This tells us that if the Universe does have one of these strange topologies, the distance to the gluing is not much smaller than the size of the whole visible Universe. If the gluing scale is greater than the size of the visible

Universe, then we would not be able to tell whether our Universe was finite or infinite, if it was flat or negatively curved.

Recently there has emerged a small piece of evidence that might be telling us that the Universe has one of these finite topologies. The observations of the effects of the Universe's past vibrations, taken by NASA's WMAP satellite, showed that the vibrations of one type were significantly depleted.[21] This is puzzling, but might be naturally explained by a finite Universe. The finite space only allows certain vibrating waves to 'fit' in. Many long waves are excluded, and as a result a deficit of long-wave vibrations would be evident in the Universe.

THE PROBLEM OF UNIFORMITY

'Somewhere over the rainbow
Way up high,
There's a land that I heard of
Once in a lullaby'

Over the Rainbow, Yip Harburg[22]

Our discussion of topology ended with the introduction of a key problem that besets cosmology: the Universe is not the same as the visible Universe (Figure 7.13). Philosophers and theologians like to talk about 'The Universe'. Unfortunately, astronomers can say little about 'The Universe'. It may be finite or it may be infinite, because all that our observations can ever gather information about is something that is necessarily finite, which we call the 'visible Universe'. The visible Universe is the region of the entire Universe from which light signals have had time to reach our telescopes since its expansion began. At present, the horizon of the visible Universe is about 42 billion light years away. Each day our visible Universe gets bigger by the distance

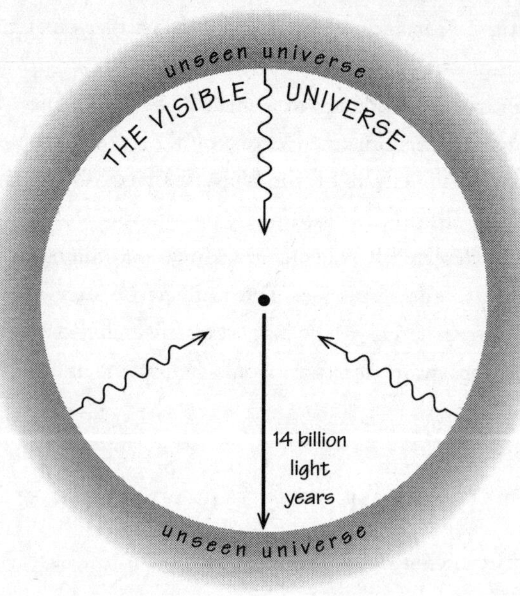

Fig 7.13 *The difference between the visible Universe and the entire Universe.*

light can travel in a day, but in practice this is not a noticeable change because we can't see right to the horizon of the visible Universe.

This distinction leads to a number of striking conclusions:

We can only see a finite fraction of the whole Universe

Regardless of how large the entire Universe is, we can only gather information about a finite part of it. The finiteness of the speed of light ensures our experience of the Universe is always a finite one.

The observable fraction is zero if the Universe is infinite in size

If the Universe is truly infinite in size, then, no matter how large the visible part of the Universe may be, it will always be an infinitesimally small fraction of the whole Universe. Unless we introduce some unverifiable assumption that our visible part of the Universe is typical of the whole, then we are always confined to gathering data about a vanishingly small part of the infinite whole.

Until fairly recently this type of argument was viewed by cosmologists as nit-pickingly philosophical and unduly pessimistic. There was no positive reason to expect the Universe to be very different beyond our horizon. However, things are now rather different. The most popular cosmological theory of the 'inflationary' Universe predicts that the Universe should be very different beyond our visible horizon, while our visible Universe should be rather smooth in its distribution of matter in stars and galaxies. If we go far enough off beyond our horizon (or wait long enough for the arrival of its distant light) we should find the Universe to be very different in its expansion, its density, its temperature, even in its laws and the number of dimensions of space and time. For the first time, the inflationary Universe provides us with a *positive* reason to expect things to be different elsewhere in the Universe.

This problem of uniformity was recognised in 1922, soon after the discovery of the first cosmological models based on Einstein's general theory of relativity, by the French mathematician Emile Borel, who wrote,

> 'It may seem rather rash indeed to draw conclusions valid for the whole universe from what we can see from the small corner to which we are confined. Who knows that the whole visible universe is not like a drop of water at the surface of the Earth? Inhabitants of that drop of water, as small relative to it as we are relative to the Milky Way, could not possibly imagine that beside the drop of water there might be a piece of iron or a living tissue in which the properties of matter are entirely different.'[23]

Borel raises a deeper problem about extrapolating from the local to the global. He suggests that the properties of matter, indeed many of the things that we regard as unalterable properties of the Universe, might differ from place to place. The inflationary Universe scenario allows this type of variation. It proposes that during the first moments of the early Universe's expansion, it came under the gravitational influence of forms of matter which were gravitationally repulsive. This accelerated the expansion dramatically, but different regions would experience slightly different periods of acceleration. The result was akin to heating up a foam of bubbles in a random way (Figure 7.14).

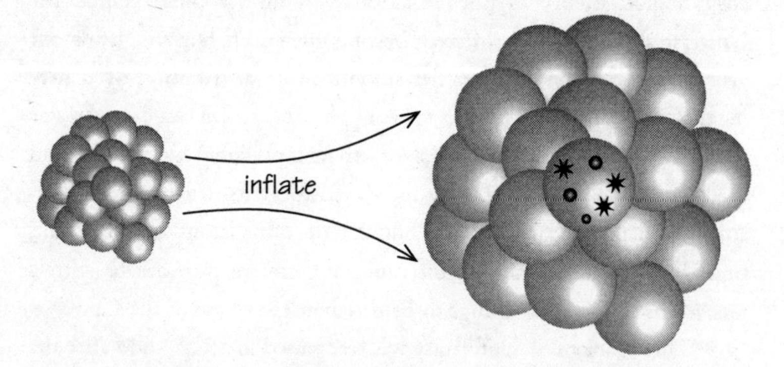

Fig 7.14 *The inflation of different parts of space. Imagine a foam of bubbles being randomly heated so that they expand by different amounts. One of the bubbles grows large enough and old enough for galaxies and stars to form. This encloses all of our visible Universe today.*

Some expand a lot, some only a little. Today we find ourselves inside one of the regions that inflated a lot – enough to allow stars and living observers enough time to evolve. Our 'bubble' is most likely to be much larger than the distance to our visible horizon today – it would be a weird coincidence if it was not. Everything in that bubble can be thought of as having the same 'genetic' code: the same physical laws and structural characteristics. Indeed, we can predict what the pattern of variations in the expansion of our visible Universe should look like from one place to another and from one direction in the sky to another. In recent years, huge amounts of effort have gone into the search for these variations in the background radiation in the Universe. It will contain the fossilised imprints of the basic quantum fluctuations that the inflation process stretches out across our bubble. Those fluctuations eventually become the galaxies and stars that populate the Universe we see today.

So far, our observations show a remarkable agreement with the predictions of the simplest inflationary universe theories. The distinctive pattern of variation that inflation imprints upon a bubble that expands for 14 billion years is there. We can see its presence imprinted on the microwave radiation that fills the Universe, a hiss of distortion from the beginning of the Universe that is even visible as interference on our television sets. If we look with instruments of exquisite sensitivity then we can pick up the micro-structure of this noise, and extract from it the information it brings about our inflationary past.

But enough of our bubble. What about the others, possibly infinite in number, that lie beyond our visible horizon? If they have inflated by different amounts, they will expand differently from our Universe, have different levels of graininess, contain different types of galaxy – possibly no galaxies at all. They will be different worlds in respect of the distribution of matter. In this sense they would be just like different places in our own world, like going from one type of geographical landscape to another as we trek across a continent. But the differences could be far more spectacular. In some versions of the inflationary universe theory the 'constants' of Nature can find themselves imprinted on different bubbles with different values. For all practical purposes, physics will be different from one bubble to another. Only in some bubbles will biochemistry and life be possible. Most remarkable of all, it turns out that even the number of dimensions of space that grow large can vary from bubble to bubble. Modern theories of the forces of Nature only seem to make good mathematical sense if there exist many more dimensions of space than the three in which we live today. The most favoured theories have ten space dimensions and we should imagine that we live on a three-dimensional surface, which is called the 'braneworld'. The three dimensions of the braneworld have, for some (as yet unknown) reason, become large while the other dimensions have stayed small – so small that they are imperceptible to us now. Many questions remain unanswered. Why did three dimensions get big? Is it hard-wired into the laws of physics that it had to be three, or is it a random process that decides this? If it's random, perhaps some bubbles have three big dimensions while others have seven or two or even none at all.

We do not know if this scenario is correct. Strenuous efforts are being made to deduce observable consequences of the extra dimensions for our three-dimensional world. The lesson we learn from unearthing these possibilities within well-developed theories of the forces of Nature is that the *observed* Universe, with its three dimensions, four forces and finite size, may not only be just a small part of an infinite whole, but an unrepresentative part at that. Its key property is that it permits complex observers like ourselves to exist within it. If that is a highly improbable property it is none the less one that we cannot fail to find that it possesses.

We can never know if the Universe is finite or infinite

If the Universe can possess the variations in space that are expected from the process of inflation, then we will never know whether the Universe as a whole is finite or infinite. Bubbles, like our own, which inflate a lot become tantalisingly close to the critical divide that separates a future of indefinite expansion from one that will end in contraction to a Big Crunch of high density. Our distance from the critical density may well be of a similar magnitude to the fluctuations in the density from region to region in the Universe. This means that we could be living in an underdense part of a much larger denser region that will eventually contract, trapping us within it: we think that we live in an infinite Universe and are going to expand forever, but we are fooled. Conversely, we may think we are living in a Universe of sub-critical density when we just inhabit an

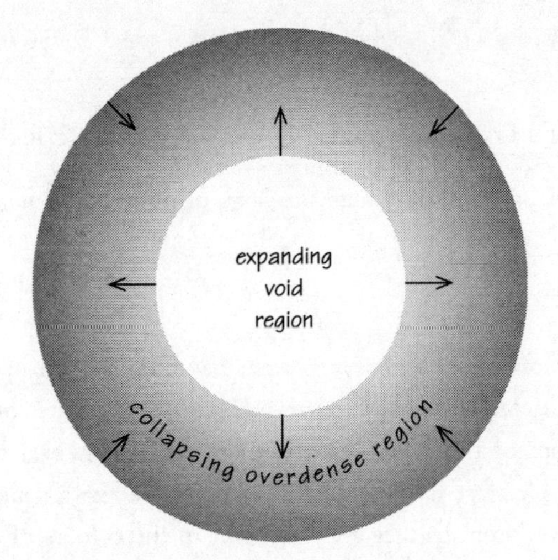

Fig 7.15 *We may think we are living in a Universe of sub-critical density when we just inhabit an underdense 'bubble' inside a Universe of super-critical density.*

underdense part of a larger bubble of super-critical density (Figure 7.15).[24] This is one of the quandaries of living life on the edge as far as universes are concerned.

A nice example is given by the decanter pictured in Figure 7.7. Around its round base it is positively curved, like the space in a closed universe that will eventually re-contract. Around the lip it is negatively curved, like a universe on course to expand forever. A civilisation of ants living near the lip of the vase would judge it to be negatively curved and might assume that the vase goes on forever. However, looking down from the third dimension we see that there are other parts of the vase where the curvature is positive. They outweigh the negative regions and the whole surface of the vase is finite.

THE PROBLEM OF ACCELERATION

'But I canna change the laws of physics, Captain!'

Scotty, Chief Engineer, *USS Enterprise*

In the last few years a new ingredient has been thrown into the mix that will determine whether the Universe is finite or infinite. Observations of the fading light from exploding stars near the edge of the visible Universe have enabled us to determine with some certainty how far away from us they are. The shift in the colours of their light enables us to measure how fast they are moving away from us. Putting these deductions together we can use them to track the expansion of the Universe.

The expected course for the expansion of the Universe was that the expansion always decelerates, because after the expansion begins there is only gravity to slow it down. No other force plays a role. But astronomers have discovered that objects near the edge of the visible Universe are accelerating away from each other. This means that the Universe must now be dominated by a mysterious new form of dark energy which anti-gravitates, repelling rather than attracting other forms of matter. Seventy per cent of the energy in the Universe needs to be in this strange form in order to explain the acceleration that we see. It provides us with a simple picture of what the Universe is made of. Ironically, the less we know about a particular part of the Universe's make up, the more abundant it seems to be!

This discovery of an accelerating Universe is a huge challenge to cosmologists. It means that the expanding Universe is likely to be on a trajectory like that shown in Figure 7.16.

The Universe appears to have been expanding for 13.7 billion years. For the first 379,000 years of that history the expansion was

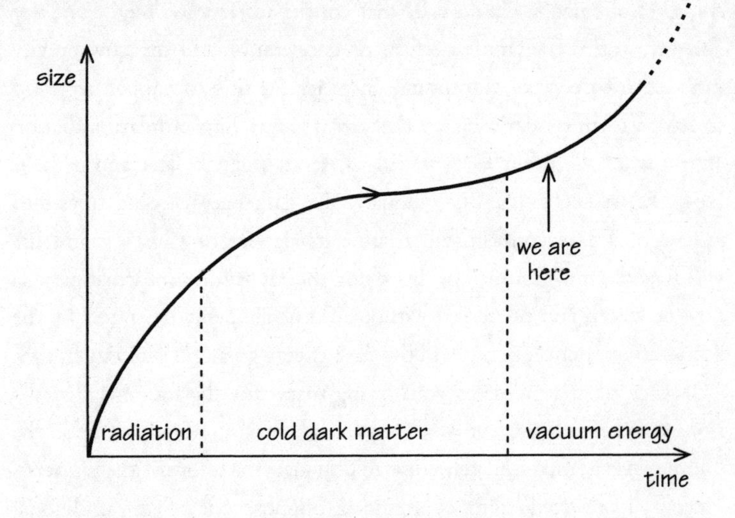

Fig 7.16 *The variation of the separation of distant parts of the universe versus time in a universe that changes from deceleration to acceleration. Our location in this cosmic history is shown.*

controlled by the gravitational pull of radiation, slowly decelerating the expansion. Then, the atoms and molecules took over, continuing to decelerate the expansion while the complicated processes that formed the stars and planets drew together the threads needed to create the cosmic tapestry we see today. After 8 billion years had passed the dark energy took over. Deceleration soon changed to acceleration and no more galaxies could form. The expansion was just moving too fast for matter to resist.

This expansion looks unstoppable. It will propel the Universe into an ever-expanding future where all forms of life, no matter how complex or advanced, appear doomed to extinction. No local structures will be able to persist and no new stars and galaxies will ever be able to form. Acceleration to infinity sounds exciting, but it marks the end of everything that we value.

Faced with such a bleak and infinite future, it is natural to seek some get-out clause. Could it be that cosmic acceleration might one day go away, so that the familiar effects of deceleration and attractive gravity can return to control the future? In order for this to happen we need to have a form of dark energy that decays away into ordinary radiation after a while. Before it fades it will accelerate our Universe and make it look like we see today. But gradually the dark energy could turn into radiation. The expansion will resume its decelerating history and life will have a faint glimmer of hope for the far future. Information can now be stored and processed without ultimately being destroyed by the accelerating expansion. But will the dark energy go away? Nobody knows.

One of the problems with trying to predict the future is that the slowest and most innocent of changes, while making no impact upon the Universe today, can ultimately come to dominate the future of the Universe, leaving all our predictions confounded. Suppose one of the traditional 'constants' of Nature were to be slowly changing with time so that the intrinsic strength of gravity or the electromagnetic force that binds atoms together were to be slowly fading. Ultimately, the cumulative effects would be overwhelming. A weakening of gravity would prevent structures like stars and galaxies persisting. A similar change in the strength of atomic forces would make the existence of atoms and stars impossible. Slow and subtle changes of this sort would eventually have a crucial role to play in fashioning the nature of asymptotia.

WHERE DOES THIS LEAVE US?

'Prediction is very difficult, especially about the future'

Niels Bohr[25]

We seem to find ourselves in a universe that will keep on expanding forever. It may accelerate faster and faster into the far, far future, or it may one day resume its gradual deceleration. Either way, the future looks infinite in time unless new things bring about unexpected changes to the future. Yet, even if the future of time is everlasting, this does not mean that space need be unending. It may be, if the topology of the Universe is simple. But it may not be. There are finite universes that expand forever. One day we may be able to determine if the topology of our Universe is of this unusual sort. Its finiteness would have constraining effects on the longest waves of radiation in space, suppressing them at their extremes so that they can fit in the space. They will leave a special signature of the shape of space which we may be able to pin down. Yet, even if we crack this problem, we still have to face the spectre of a Universe that is slightly different from place to place. Locally it may have the characteristic of a universe destined to keep on expanding forever and make you think that it is infinite. But, if we could only look over the horizon, we might see a denser hinterland that is enough to curl up the geometry of space into finiteness. Remarkably, our visible Universe expands so close to the dividing line that separates indefinite expansion from eventual contraction – closer than 2 per cent – that very small variations in the density of the Universe from one region to another can easily bring about this dilemma.

Our Universe may be infinite, but this is one of its most closely guarded secrets. Infinity is guarded by finiteness. It is an attribute given perhaps to universes, but protected by the limits that exist on how fast information can spread. You can discover whether the Universe is infinite, but the learning will take an infinite time.

THE SHINING

'Day and night
Night and day'

Night and Day, Cole Porter

Fig 7.17 *Looking into the woods. Everywhere your line of sight ends on a tree trunk. We should see a forest of stars if we look out into the universe.*

Edmond Halley (1656–1742) is known throughout the world because of the comet that bears his name. Halley calculated its orbit and determined that comets seen in 1531, 1607 and 1682 were the same object that followed a 76-year orbit (on average).[26] Unfortunately, Halley died in 1742, and never lived to see his prediction come true when the comet returned on Christmas Eve in 1758. Its return is rarely spectacular, but it is one of those events that links the generations. I saw it last time in 1986; it passed us by as we watched from a garden in Lewes, Sussex, but my friend and distinguished colleague Bill McCrea[27] trumped us all by telling us he'd seen it the time before as well!

Predicting the comet's return is not the most important piece of astronomy that Halley did. He ought to be known for noticing something much more interesting than a comet, but unfortunately, by a historical accident the credit went to someone else. Halley was interested in the possibility that the Universe might be infinite. It was a rather fashionable question in his day: a nice point of confrontation between astronomical reality and the philosophical picture one might have of the Universe as a whole.

Halley noticed a simple but profound problem that seemed to confront an infinite universe:

> 'I have heard urged that if the number of Fixed Stars were more than finite, the whole superficies of their apparent Sphere would be luminous.'[28]

Indeed, it would even be a problem for a finite universe if it was big enough. As we have noted before, look out into the woods and what do you see? Everywhere your line of sight seems to end on the trunk of a tree (Figure 7.17).

Halley realised that a universe containing an infinite number of stars creates exactly the same situation. If we look out into the sky we should find that every line of sight ends on the surface of a star. The result: the entire sky should shine like the surface of a star, day and night. But it doesn't. And that is Halley's Paradox – except that it's called Olbers' Paradox![29]

There are some simple ways to avoid the paradox. Suppose that space goes on forever but the stars do not. Space is infinite but the material universe is finite. In this case there would be only a finite amount of light reaching us from the stars and our skies might grow dark when the Sun sets (see Figure 7.18).

This seems an artificial solution. If the Universe has infinite spatial extent, why is there a little finite pocket of stars and planets somewhere within it? What makes that place special? We are forced to adopt a perspective that is anti-Copernican in the sense that it gives us a special place within the infinite Universe. Indeed, why bother with the infinite space beyond the finite distribution of matter? It is superfluous to this

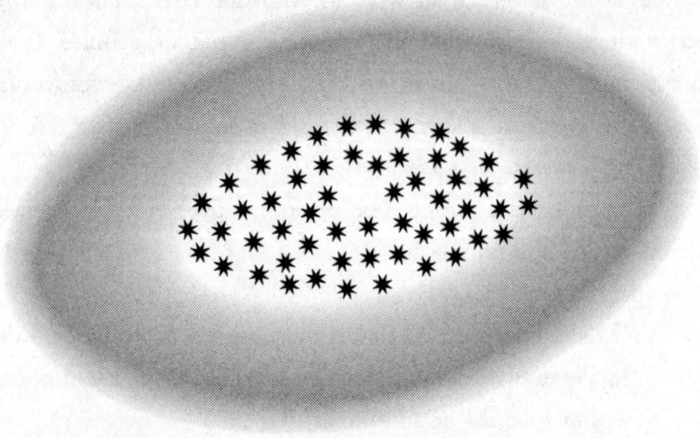

Fig 7.18 *An infinite universe that only contains a finite cluster of stars around us and nothing beyond because the stars have all died or never existed there.*

model universe and seems to exist only so that the picture of an infinite space can be accommodated.

Another option is to stick with the infinite Universe, but suppose that the stars within it – which may be infinite in number – came into being a finite time ago. There has only been time for light to reach us from those stars that are closer than the distance that light can have travelled since the stars turned on. In effect, the finiteness of the speed of light makes the Universe seem finite. So long as the density of stars dotted around space is not too great, it would be possible for the night sky to remain dark.

The English poet Edward Young (1683–1765), a contemporary of Halley, composed a poem about the nature of the Universe, entitled *Night Thoughts,* that contemplates the impact of the speed of light on our reception of light from far away in an infinite universe. He tells us that

'So distant (says the sage) 'twere not absurd
To doubt if beams, sent out at nature's birth,
Are yet arrived at this so foreign world;
Though nothing half so rapid as their flight.'[30]

During the 300 years since Halley noticed this strange aspect of the Universe there have been many attempts to resolve the paradox of why the sky is dark at night.[31] Some suggested that interstellar dust might obscure the light from many of the stars, but it was soon realised that such a trick doesn't help at all. The intervening dust just heats up and eventually radiates the same radiation energy that it absorbs. A full understanding of the darkness had to wait until Einstein's theory of general relativity and the discovery that the Universe is expanding.

In order that our Universe can contain life, it must contain elements heavier than simple hydrogen and helium. These heavier biochemical elements, like carbon, nitrogen, and oxygen, are made in the stars. Nuclear reactions inside the stars slowly burn hydrogen to helium and helium to beryllium, carbon and oxygen. These complex elements are spread round the Universe when stars explode and die. Eventually, they find their way into planets and people, but this is a long slow process. It takes billions of years to complete the stellar alchemy needed to produce carbon. Thus we begin to see why the Universe is so big and so old. The Universe must be billions of years old in order to produce the basic building blocks of life. If it is also expanding, then it must be billions of light years in size. A universe trimmed down to be only as big as our own Milky Way galaxy, with its 100 billion stars, sounds plenty big enough, but it would be little more than a month old – not enough time to produce the building blocks of living complexity. It would contain no observers. We should not be surprised to find that our Universe is so large. We couldn't exist in one that was significantly smaller.

The huge age needed if the Universe is to be able to evolve life has other consequences too. The expansion dilutes the density of matter, making the average distance between atoms, stars and planets increasingly large. It also lowers the temperature of the radiation in the Universe, so that today it is only 2.7 degrees above absolute zero. The coldness and comparative emptiness of the Universe are just inevitable by-products of the great age required of a life-supporting universe. Ironically, these features of the Universe appear antithetical to life, yet they are necessary features of any expanding universe that contains the building blocks needed to produce living complexity.

It is the inevitability of the low density of matter in a universe that has expanded for long enough to produce the building blocks of life that provides us with a resolution of Halley's paradox. The night sky is dark because the Universe does not contain enough matter and energy to make it bright. If we were suddenly to transform all the matter in the Universe into radiation, we would hardly notice: the temperature of the radiation in the Universe would simply rise from 2.7 to about 10 degrees above absolute zero. The great age needed if a universe is to be a habitat for complex atom-based life means large size. This means that the density of matter and energy of radiation is inevitably degraded to be negligible when the universe is old enough for observers to see it. We should not be surprised that the sky is dark at night. We could not exist in a universe where it was bright.

Halley's background glare of radiation from the background universe did exist in the first moments of the expansion, 13 billion years ago, but the expansion has degraded its energy so that now it is a tiny crackle of microwave noise. This modern resolution of Halley's paradox means that it does not have anything to tell us about the finiteness of the Universe. Expanding universes with a dark night sky can be finite or infinite.

chapter eight

The Infinite Replication Paradox

'The sane person prides himself on his ability to be unaffected by important facts, and interested in unimportant ones. He refers to this as having a sense of perspective, or keeping things "in proportion".'

Celia Green[1]

A UNIVERSE WHERE NOTHING IS ORIGINAL

'You can do everything right, strictly according to procedure, on the ocean, and it'll still kill you, but if you're a good navigator at least you'll know where you were when you died.'

Justin Scott[2]

Imagine living in a universe where nothing is original. Everything is a fake. No ideas are ever new. There is no novelty, no originality. Nothing is ever done for the first time and nothing will ever be done for the last time. Nothing is unique. Everyone possesses not just one double but an unlimited number of them.

This unusual state of affairs exists if the universe is infinite in spatial extent (volume) and the probability that life can develop is not

Fig 8.1 *The infinite replication paradox brought dramatically to life in a universe where nothing is original in the Milan production of* Infinities, *directed by Luca Ronconi.*

equal to zero. It occurs because of the remarkable way in which infinity is quite different from any large finite number, no matter how large the number might be.[3]

In a universe of infinite size, anything that has a non-zero probability of occurring must occur infinitely often. Thus at any instant of time – for example, the present moment – there must be an infinite number of identical copies of each of us doing precisely what each of us is now doing. There are also infinite numbers of identical copies of each one of us doing something other than what we are doing at this moment. Indeed, an infinite number of copies of each of us could be found at this moment doing anything that it was possible for us to do with a non-zero probability at this moment (Figure 8.1).

It is widely believed that the replication paradox was first discussed explicitly by the German philosopher Friedrich Nietzsche in *The Will to Strength* (1886).[4] He realises that

> 'the universe must go through a calculable number of combinations in the great game of chance which constitutes its existence . . . In infinity, at some moment or other, every possible combination must once have been realized; not only this, but it must also have been realized an infinite number of times.'[5]

Yet, Nietzsche himself writes that he first learnt of the idea in the writings of the German poet and essayist, Heinrich Heine (1797–1856). In one of Heine's works he had read a version of this argument where the infinite recurrence relies on the eternity of time rather than the infinity of space:

> 'time is infinite, but the things in time, the concrete bodies are finite . . . Now, however long a time may pass, according to the eternal laws governing the combinations of this eternal play of repetition, all configurations that have previously existed on this earth must yet meet, attract, repulse, kiss, and corrupt each other again . . . And thus it will happen one day that a man will be born again, just like me, and a woman will be born, just like Mary.'[6]

The spatial replication paradox has all sorts of odd consequences aside from the psychological unease that it creates. We believe that the evolution of life is possible with non-zero probability because it has happened on Earth by natural means. Hence, in an infinite universe there must exist an infinite number of living civilisations. Within them will exist copies of ourselves of all possible ages. When each of us dies, there will always exist elsewhere an infinite number of copies of ourselves, possessing all the same memories and experiences of our past lives but who will live on to the future. This succession will continue indefinitely into the future and so in some sense each of us 'lives' forever.

This argument has entered theological discussion in a provocative way as well. For suppose that we apply the same reasoning to the crucifixion of Christ. If it has a finite probability of occurring, then it must have occurred infinitely often elsewhere already in an infinitely large Universe. This argument was used by St Augustine to claim that life must be unique to the Earth or the crucifixion would have to have occurred on other worlds as well. Thomas Paine argued that the existence of life elsewhere was obviously true, and therefore the crucifixion did not occur (or at least could not have had its claimed effects).

We could ask what might happen if we were to meet one of our copies. One might think this was just like shadow boxing in the mirror, but there is no reason to think that our single double would act as we do. We may have identical pasts up until this moment, but confronted with a new situation we might well respond differently, just as two identical twins might do. In the future our experiences and choices would become increasingly different. Indeed, it is more probable that our futures would diverge than stay similar. Yet, elsewhere in the infinite universe there would have to be copies of each of us making the same decisions, and being in every respect identical. It is as if every possible decision that we could have taken at every moment is taken because there is always someone, somewhere, who lives a past life identical to our own, but who then takes one of the options I didn't take about what to do next. This idea was mentioned by Herbert Spencer in 1896 in respect of the course of the evolutionary process:

> 'And thus there is suggested the conception of a past, during which there have been successive Evolutions analogous to that which is now going on; and a future during which successive other such Evolutions may go on – ever the same in principle but never the same in concrete results.'[7]

This is reminiscent of the debate that goes on still, about whether the evolutionary process would have unfolded in a completely different way if it had been slightly perturbed at some very early stage.[8]

One of the curious features of this 'theory' is that if it is true, it cannot be original. It has already been proposed infinitely often in the past.

THE GREAT ESCAPE

> 'If one were to believe the Pythagoreans, with the result that the same individual things will recur, then I shall be talking to you again sitting as you are now, with this pointer in my hand, and everything else will be just as it is now.'
>
> Eudemus of Rhodes[9]

This conclusion, that we are surrounded by an infinity of clones, is so peculiar and worrying that we should ask if there is any way to avoid it. The simplest escape clause is to maintain that the Universe is finite. Some cosmologists find the infinite replication paradox so unsavoury that a finite Universe would be a welcome escape from its implications. Others are placated by the realisation that the finite speed of light ensures that only a finite part of an infinite universe can ever be seen by us. In an infinite universe with the contents of our own, we would have to travel about 10^N metres, where $N=10^{27}$, before we encountered our first double with near certainty.[10] This is a gigantic distance.[11] The distance out to which we can see with perfect telescopes, which defines the size of the visible Universe, is only about 10^{27} metres (see Figure 8.2). In the far future our visible Universe will extend far enough to include our doubles, but by then we will be long dead and the Universe will be too old to contain any stars and solar systems. If we go out to a distance of 10^N metres, where $N=10^{119}$, then we will encounter regions of the size of our entire visible Universe today which are identical to it.

Another way of avoiding the conclusion of infinite replication in an infinite Universe is to maintain that the probability of life developing in the Universe is *zero*. If so, then the number of replicas of each of us existing in the Universe at this moment will be $0 \times \infty$, which can equal any finite number because if we divide 1 by 0 we get infinity, if we divide 2 by 0 we get infinity, and so on. In this case there might be only *one* copy of you elsewhere but, equally, there might be *one million billion* of them. The idea that life comes into being despite having zero probability of doing so naturally is like saying that it has a miraculous or supernatural origin.[12] If life is pre-programmed only to evolve on planet Earth then the paradox is avoided.

Another possible escape route from the paradox is to imagine that there might be an infinite number of possible life-forms. George Ellis and Geoff Brundrit consider and reject this possibility:

The argument for identical beings existing would not hold if there were an infinity of different possible life-forms, but we believe that this can be argued against on the basis that there are only a finite number of elements and there is a maximum size for stable molecules; so the number of molecular configurations on which other life-forms could be based is finite. That is, the kind of life-forms we experience are almost certainly a finite (non-zero) fraction of all possible life-forms. Neither do we believe that a simple use of the indeterminacy principle [of Heisenberg] will solve the problem; for while it is true that . . . quantum theory allows

Size of the Earth	1.28×10^{7} metres
Distance to the Sun	1.5×10^{11} metres
Distance to the nearest star	6×10^{16} metres
Distance to the edge of our Galaxy	3×10^{19} metres
Distance to the edge of the visible Universe	10^{27} metres
Distance to the first copy of you or me	$2 \times 2^{5 \times 10^{28}}$ metres $\approx 10^{10^{28}}$ metres
Distance to the first copy of the Earth	$10^{7} \times 2^{10^{51}}$ metres $\approx 10^{10^{50}}$ metres
Distance to the first copy of the entire visible Universe	$10^{27} \times 2^{10^{120}}$ metres $\approx 10^{10^{119}}$ metres

Fig 8.2 *How far do you need to go before you find a duplicate Earth or meet your exact double? In the volume of our visible universe of diameter 10^{27} metres there is space for $N = 10^{120}$ subatomic particles and so the state of the universe (with each of these particle-sized volumes either occupied or not) has 2^N distinct possible configurations. We would expect to have to travel about $2^N \times 10^{27}$ metres before running into duplicates of our visible universe.*

an infinity of later histories from any initial configuration, the time reverse is also true: any final configuration may have been arrived at from an infinity of earlier histories. Accordingly the effect of the uncertainty principle is to make more complex the transition from a . . . set of initial states to a . . . set of final states; but it does not appear thereby to set to zero the probability of occurrence of planets closely similar to the Earth . . . We can obtain non-zero probabilities for occurrence of conditions within any *specified* finite neighbourhood of those on Earth. Consequently environments may be expected which may – if we make these neighbourhoods small enough – be plausibly argued to lead to very similar *actual* histories.'[13]

THE TEMPORAL VERSION – BEEN THERE, DONE THAT

'The thing that hath been, it is that which shall be; and that which is done is that which shall be done: and there is no new thing under the sun.'

Ecclesiastes[14]

The myth of the 'eternal return' is an ancient one. We find it in Eastern and Western philosophies in many guises. The world is seen as an eternal process of which we are transient parts, destined to be replaced by reincarnated replacements who are identical, similar, or even quite different, according to which version of this meta history you choose to believe. It has been argued that one of the important contributions that the Judaeo-Christian tradition made in shaping our world-view into a form where progress was desirable was to subdue this cyclic picture of history. The linear picture of history, with a beginning, and a future that is different from the present, gives a rationale for scientific investigation of the world and a basis for social progress and ethics.

In modern times there have been scientific cosmologies which share some of the features of the ancient cyclic universes. It has been suggested that the expanding universe we witness is but one cycle of a continual oscillatory process of expansion and contraction (Figure 8.3).

Then there have been steady-state universes, in which the average appearance of the universe is always the same. It expands always at the same rate. These universes look the same on the average at whatever time in history they are observed. Once they were viable descriptions of our Universe, but observations of the background radiation and lightest elements show that the Universe was once hotter and denser than it is today. It is not in a local steady state. But if the universe had been unevolving like that, there would have been some unusual consequences. Here is one.

If the universe is in a steady state, so that it has the same structure on the average at all times, and is infinitely old, then the infinite replication paradox also operates in *time* as well as in space. Anything that has a finite chance of occurring will have occurred infinitely often in past history. No idea can be new. Such universes have a remarkable feature.[15] Because there is a finite chance of intelligent life evolving, it must be infinitely common and as time goes on there should be a huge proliferation in the frequency of living beings. We should expect to see them routinely. It means that if there is a non-zero probability of intelligent life developing in an infinitely old universe, then it will have already done so infinitely often. There should be ETs everywhere. Again, there is a paradox. We do not see ETs everywhere. That doesn't exclude the possibility that ET is very small – nanoscopic – and actually rather prevalent. Astronomers like to come up with arguments that reconcile the high probability of ETs with the complete lack of any observational sign that they are out there.[16]

Today, there is a popular picture of the Universe which views it as a steady-state process when viewed from an infinite scale.[17] It is like a never-ending sea of expanding and contracting bubbles which continually spawn further bubbles, ad infinitum. As we shall see in the next chapter, this picture, the so called eternal inflationary universe, seems to require neither beginning nor end. In each of the bubbles there is the possibility that aspects of the laws of physics and the values of the constants of Nature can fall out differently. Yet, the infinite replication paradox applied to these bubble worlds is complicated by the fact that there may well now exist an infinite number of different forms of living complexity. If there are, then it is not required that we have an infinite number of replicas in cosmic history within the other worlds. But if the number of ways of making living complexity is finite, then our replicas again stretch out endlessly through space and time.

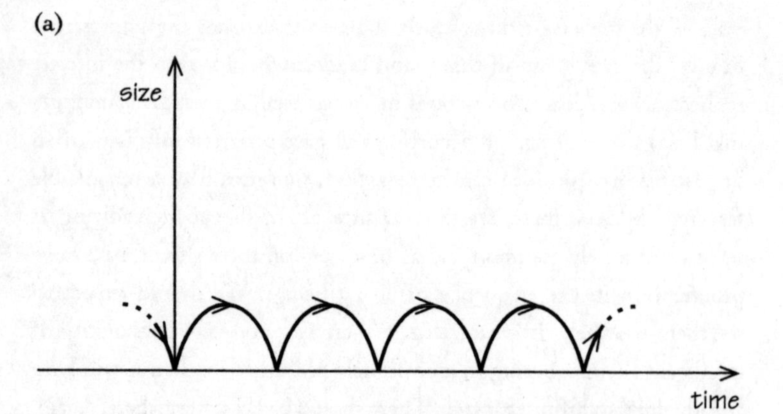

Fig 8.3 *Three possible futures for an oscillating universe: (a) An oscillating universe of many equal cycles. (b) The size of the expansion maximum should grow as the oscillations continue because of the second law of thermodynamics. (c) If a dark energy exists which accelerates the expansion then an increasing sequence of cycles must eventually end and be replaced by a state of indefinite expansion.*

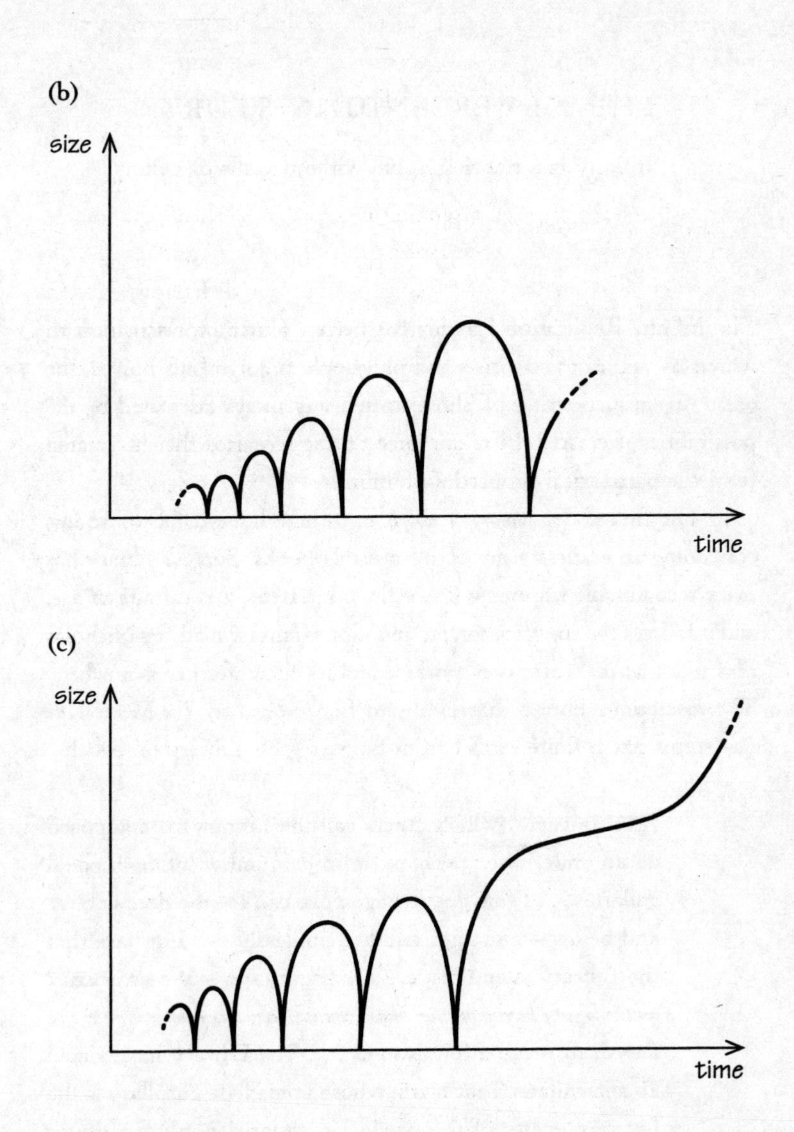
(b)
size
time
(c)
size
time

THE NEVER-ENDING STORY

'Infinity is a floorless room without walls or ceiling.'

Anonymous[18]

The Infinite Replication Paradox has been a source of fascination to writers as well as to scientists and philosophers. Jorge Luis Borges, the great Argentinian writer of short stories, was always fascinated by the possibilities it created. Here are three of the scenarios that he created from the paradoxical properties of infinity.

The first is *The Library of Babel*, an infinite honeycomb of rooms containing an endless array of all possible books. Borges's library has many recognisable features – it is infinite in extent, it is infinite in age, and it follows the specification for an infinite universe made by Nicholas of Cusa: that its centre is everywhere and its circumference is nowhere. The mysterious library that Umberto Eco evoked in *The Name of the Rose* seems like a finite extract from Borges's great Library of Babel:

> 'The universe (which others call the Library) is composed of an indefinite, perhaps infinite number of hexagonal galleries . . . From any hexagon one can see the floors above and below – one after another, endlessly . . . I declare that the Library is endless . . . *The Library is a sphere whose exact centre is any hexagon and whose circumference is unattainable* . . . I wish to recall a few axioms . . . The Library has existed ab aeternitate. That truth, whose immediate corollary is the future eternity of the world, no rational mind can doubt . . . Some five-hundred years ago . . . [a] philosopher observed that all books, however different from one another they might be, consist of identical elements: the space, the

> period, the comma, and the twenty-two letters of the alphabet. He also posited a fact which all travellers have since confirmed: *In all the Library, there are no two identical books*. From these incontrovertible premises, the librarian deduced that the Library is 'total' – perfect, complete, and whole – and that its bookshelves contain all possible combinations of the twenty-two symbols (a number which, though unimaginably vast, is not infinite) – that is, all that is able to be expressed, in every language. *All* – the detailed history of the future, the autobiographies of the archangels, the faithful catalog of the Library, thousands and thousands of false catalogs, the proof of the falsity of those false catalogs, a proof of the falsity of the *true* catalog, the Gnostic gospel of Basilides, the commentary upon that gospel, the commentary on the commentary on that gospel, the true story of your death, the translation of every book into every language, the interpolations of every book into all books, the treatise Bede could have written (but did not) on the mythology of the Saxon people, the lost books of Tacitus.'

But the number of possible permutations of a finite number of letters is not infinite. The inventory of the library could only be made infinite by exploiting its infinite age to fill it with books of all possible lengths. Borges senses that he is running into Bertrand Russell's paradox of the set of all sets that are not members of themselves – is it a member of itself?

> 'When it was announced the library contained all books, the first reaction was unbounded joy . . . There was no problem whose eloquent solution did not exist – somewhere in some hexagon. On some shelf in some hexagon, it was argued, there must exist a book that is the cipher and

> perfect compendium *of all other books*, and some librarian must have examined that book; this librarian is analogous to a god.'

Yet, Borges senses that the scenario he has created does not quite work and steps back to imagine that his Library is without boundary rather than infinite. He suggests that it achieves this by periodic wrapping around in space:

> 'I have just written the word "infinite". I have not included that adjective out of mere rhetorical habit; I hereby state that it is not illogical to think that the world is infinite . . . those who picture the world as unlimited forget that the number of possible books is *not*. I will be bold enough to suggest this solution to the ancient problem: *The Library is unlimited but periodic*. If an eternal traveller should journey in any direction, he would find after untold centuries that the same volumes are repeated in the same disorder – which repeated, becomes order: the Order.'

Thus in the end the Library contains only a finite number of different books, but a reader would never come to the end of the shelving.

Borges returned to the dilemma in *The Garden of Forking Paths*. Here a story bifurcates like a forking path. All possible decisions are taken and lead to different histories. This is a scenario reminiscent of the 'many worlds' interpretation of quantum mechanics, in which all possible histories actually occur.

> 'Before unearthing this letter, I had wondered how a book could be infinite. The only way I could surmise was that it be a cyclical, or circular, volume, a volume whose last page would be identical to the first, so that one might go on indefinitely. I also recalled the night at the centre of the *1001*

Nights, when the queen Scheherazade (through some magical distractedness on the part of the copyist) begins to retell, verbatim, the story of the 1001 Nights, with the risk of returning once again to the night on which she is telling it – and so on *ad infinitum* . . . Almost instantly, I saw it – the garden of forking paths was the chaotic novel; the phrase 'several futures (not all)' suggested to me the image of a forking in *time* rather than in space . . . each time a man meets diverse alternatives, he chooses one and eliminates the others; in the work of . . . Ts'ui Pen, the character chooses – simultaneously – all of them. *He creates*, thereby, 'several futures', several *times*, which themselves proliferate and fork . . . a stranger knocks at [Fang's] door . . . Naturally, there are various possible outcomes – Fang can kill the intruder, the intruder can kill Fang, they can both live, they can both be killed, and so on. In Ts'ui Pen's novel, *all* the outcomes in fact occur; each is the starting point for further bifurcations. Once in a while, the paths of that labyrinth converge: for example, you come to this house, but in one of the possible pasts you are my enemy, in another my friend . . .

That fabric of times that approach one another, fork, are snipped off, or are simply unknown for centuries, contains *all* possibilities. In most of those times, we do not exist; in some, you exist but I do not; in others I do and you do not; in others still, we both do. In this one, which the favouring hand of chance has dealt me, you have come to my home; in another, when you come through my garden you find me dead; in another, I say these same words, but I am an error, a ghost . . . Time forks, perpetually, into countless futures. In one of them, I am your enemy.'

Borges returns to the paradoxes of the infinite for the last time in his story *The Book of Sand*, where a man comes into possession of a fabulous book with an infinite number of pages. It contains everything, but once the owner had turned a page he could never find it again. For it is a property of Cantor's uncountable infinities that between any two numbers with unending decimals there are an infinite number of other numbers. Borges begins

> 'The line is made up of an infinite number of points; the plane of an infinite number of lines; the volume of an infinite number of planes; the hypervolume of an infinite number of volumes . . . No, unquestionably this is not . . . the best way of beginning my story.'

The strange book has no first page and no last. The book became to its owner 'a nightmarish object, an obscene thing that affronted and tainted reality itself'. But simply to burn it seemed to him an action fraught with danger. Might not the smoke, if not the consequences, be infinite? Better to hide it; and where better to hide than in a crowd. Slipping back to the National Library where he worked, trying not to notice where his hand was reaching, he slid the *Book of Sand* 'on one of the basement's musty shelves'.

THE ETHICS OF THE INFINITE

> 'There is a concept which corrupts and upsets all others. I refer not to Evil, whose limited realm is that of ethics; I refer to the infinite.'
>
> Jorge Luis Borges[19]

John Chapman was the son of English immigrants to the United States from Yorkshire. He was born in Leominster in Massachusetts in 1774 and died seventy-one years later near Fort Wayne in Indiana. In between he lived a life that might have impressed St Francis of Assisi had he been transported to the great American frontier. Chapman was known in his time as the Apple Tree Man, or simply 'Johnny Appleseed'. For nearly fifty years he roamed the north-west territories helping people, tending animals and protecting the environment.[20] He dispensed advice to the new settlers and indigenous Indian communities alike and was warmly received by them all, recognised by the immigrants as a friend and philanthropist, admired by the Indians as one touched by the Great Spirit. Living with great simplicity and eating no meat, he was ahead of his time. Yet, he was no other-worldly dreamer. Appleseed bought and sold his plots of land and his trees with care, picking the best sites and accumulating significant funds to keep on extending his planting and philanthropic programmes. Today he would undoubtedly be the president of a large charitable foundation.

Chapman began his great tree-planting project in 1800 when he was twenty-six years old. Moving ahead of the waves of westward-moving immigrants, he embarked upon a programme of planting apple trees here, there, and everywhere. Yet, his actions were carefully planned and his nurseries were carefully protected from the foraging wild deer by logs and bushes or fences. He would return periodically to prune the trees, repair the fences and sell the plots of trees to any who wanted them, and many small towns eventually grew up around these appleseed plantations. Those who could not afford to buy the trees or plots of land never went away empty handed. A payment of used clothes, old shoes, or an I-O-U that he never redeemed, was enough. In addition to apple trees he also planted medicinal herbs, and for a long period the herb fennel was known as 'johnny weed' as a small tribute to the effectiveness of his planting programmes.

So what does Johnny Appleseed have to do with infinity? All his projects were finite, his seeds were finite in number, and his actions did a finite amount of good. His drive to plant more trees was understandable. More trees mean more apples, more apples mean more food, more food means less hunger and better health for more people. Every new seedling planted means that this good is multiplied and this is an important imperative to good works. Doing more of them adds to the good that has already been done.

Johnny Appleseed's approach to good works is one that underlies the ethical systems that are contained in many religions and the codes of behaviour that guide those who claim to have no religion at all. It makes simple sense in a finite Universe. But there are curious problems if the Universe is infinite. Faced with an infinite Universe that contains an infinite number of apple trees already, Johnny's impact is blunted. Add another apple tree and there are still an infinite number of trees. Nothing he can do can increase the number of apple trees. He can move them around so as to increase the local density of trees in our neighbourhood, but no amount of planting can add to the total apple-tree content in the Universe.

This neutralisation of apple-tree planting is superficially not so worrying, but it has more profound analogues elsewhere. If the total amount of good (or evil) in the Universe is infinite, then nothing we can do (or fail to do) can add to it: infinity plus anything is still just infinite.[21] This is the first ethical dilemma of an infinite Universe. If ethical imperatives are based simply upon doing more good things, then they make no sense in an infinite Universe. To escape this problem we need to think about other imperatives for right actions.

There are deeper ethical problems waiting for many of these alternatives as well. The problems all flow from the basic paradox of infinite replication that afflicts any infinite universe: if there is a finite chance of something happening then it is happening elsewhere infinitely often at this moment. This implies that there are infinitely many copies of each of us elsewhere taking the same choices that we are taking at this moment. But there are also infinitely many copies of each of us taking all of the other possible choices that we could have taken as well.

This pathology of infinite worlds has serious implications. Why should we act to prevent evil if there are infinitely many copies of ourselves where the evil alternatives are taken? There would seem to be worlds where Hitler prevailed and where evil overcomes good always, just as there are worlds where good overcomes evil. Nor are the concerns about these conclusions only ethical. Some altruistic actions are found to be optimal strategies for individuals to coexist with others and can be regarded therefore as consequences of evolution by natural selection. The existence of an infinite number of communities of living creatures who do not act according to an evolutionarily stable strategy, and yet survive, undermines our ability to understand our world in such terms.

If we pick on particular guides to ethical action, for example Kant's adoption of the Biblical instruction to treat one's neighbour as oneself (assuming he or she has the same tastes) in the form 'do as you would be done by', then similar problems ensue. At about the same time that Johnny Appleseed was busy sowing the seeds of apple pies in the future, Friedrich Nietzsche was arguing that we should act as if we knew our actions would be infinitely repeated. Deeply impressed by the infinite replication paradox and its implications that he first highlighted, he was thinking about a cyclic Universe that repeats itself in time forever, so that eventually each of us will be reborn and repeat the same actions again, with the same effects (as well as all others) infinitely often. If something you do today will be repeated infinitely often, he thinks that this should encourage us to act for the good. But this imperative is undermined by the knowledge that we will also be reborn to perform all possible actions infinitely often as well, even the bad ones that are the antithesis of the good ones he hopes to encourage.

For some religions, these paradoxes of the infinite appear to have unsatisfactory, even unacceptable, consequences. If our world is marred by evil and needs redemption and transformation, then what of worlds elsewhere? If by worlds we mean civilisations similar to our own, on planets dotted around an infinite Universe, then there must be infinitely many of them that did not undergo a fall from grace and who don't need to be redeemed and transformed. The famous science fiction trilogy of C.S. Lewis envisages just such a scenario.[22] The Earth is a moral pariah in the Universe. It is the one place where evil has arisen and where redemption is necessary. Extraterrestrial beings are perfect and have not had to be saved from the consequences of their evil actions. The paradox of an infinite Universe makes this scenario inevitable, along with all its possible variations as well.

It is clearly important that ethics place weight on individual actions, rather than on outcomes in an infinite Universe. Otherwise why should you not kill some other person when an infinite number of identical copies of that person are living on, elsewhere in the Universe? Our actions never seem to have irrevocable consequences in an infinite Universe if we take a global view. And how do we feel about those exact replicas of ourselves that exist throughout an infinite Universe?

We have already mentioned how St Augustine worried about these problems for the Christian doctrine of the Incarnation. If there are other worlds that needed redemption, this would require the Incarnation to have happened there too. He regarded this as impossible ('Christ died once for sinners . . .') and hence concluded that these civilisations could not exist. He would also have had to conclude that the Universe could not be infinite. Interestingly, more than a thousand years later, the humanist philosopher Thomas Paine would draw the opposite conclusions from the same considerations. Regarding the existence of extraterrestrials as incontrovertible, and assuming that according to Christian doctrine they would need redeeming by the act of incarnation, he concluded that the Incarnation did not happen here or anywhere else.

If all possible worlds exist in any sense, either as sequences of events elsewhere in our infinite Universe or as other universes which may themselves be finite or infinite, there is a further theological quandary. The God's eye view becomes puzzling. God sees all possible worlds. In an infinite number of them evil arises and redemptive action is required; but in an infinite number of other worlds no evil arises and redemption is not required. But perhaps these evil-free futures are not logically possible, given the strictures of the laws of Nature as we know them or in the presence of self-conscious beings with freewill. When actions have never-ending chains of consequences, maybe it is not possible for all of them to have only 'good' consequences for other people.

These bizarre consequences of infinite universes have persuaded many that there is something morally repugnant, if not incoherent, about an infinite Universe. In practice, we are shielded from the force of the infinite sets of doppelgängers by the finiteness of the speed of light. It limits our interaction to being within a finite horizon equal in extent to the distance that light can have travelled since the Universe began. Yet, this is not enough for some. It may also be cold comfort if all possible universes displaying all possible sequences of events exist in parallel to our own Universe. What is the status of good and evil when all possible outcomes actually arise somewhere in the great universal catalogue? If our Universe does indeed possess some deep purpose or meaning, then the finiteness of the Universe begins to look like an increasingly desirable, if not essential, part of its design. The alternative is to deny the outworking of the collection of all possibilities in actuality. Maybe only a finite number are viable histories; perhaps there is a convergence in the evolutionary process that leads to only a finite number of histories that contain periods in which consciousness and ethics emerge.[23]

This chain of argument leads us towards a resolution of our dilemma that is rather like that employed to make sense of the host of life-supporting 'coincidences' about the constants of Nature and the structure of our observed Universe that appear to exist. If all possible combinations of the constants of Nature are to be found amongst the ensemble of all possible worlds, then is it not inevitable that we find ourselves inhabiting one of that relatively small number in which the constants take on values which permit living complexity to evolve and persist?[24]

Could such an approach help us out of our ethical problems with an infinite Universe? It would have to be that a host of possible worlds were defunct because they made the existence and persistence of conscious life impossible. While this is believable for the extreme cases in which all possible moral outrages are committed so that mutual assured destruction always occurs, it is not a convincing panacea. After all, many of those other worlds that display an abundance of bad behaviour look embarrassingly like parts of our own history. It is not too hard to imagine the victory of evil over good. It need not lead to extinction, merely tyranny.

There is no easy answer to the ethical problems presented by an infinite Universe. Perhaps there is something wrong with our Earth-centred view of ethics. Or perhaps the uniqueness of mind is more deeply woven into the woof of reality and the Universe must be finite. There is one haunting feature of our Universe that may one day reveal more of the mystery of replication. Matter seems to be composed of populations of identical elementary particles. We give them names like electrons, neutrinos, and quarks. Once you have seen one electron you have seen them all. What is it that guarantees their identical character? Does it matter whether they are infinite in number?

chapter nine

Worlds Without End

'And since space is divisible in infinitum, and Matter is not necessarily in all places, it may be also allow'd that God is able to create Particles of Matter of several Sizes and Figures, and in several Proportions of Space, and perhaps of different Densities and Forces, and thereby to vary the Laws of Nature, and make Worlds of several sorts in several parts of the Universe.'

Isaac Newton[1]

OTHER WORLDS IN HISTORY

'"But do you really mean, sir", said Peter, "that there could be other worlds – all over the place, just around the corner – like that?"

"Nothing is more probable", said the Professor, taking off his spectacles and beginning to polish them, while he muttered to himself, "I wonder what they *do* teach them at these schools."'

C.S. Lewis, *The Lion, The Witch and The Wardrobe*[2]

Human beings have never been satisfied with one world. At first they wanted to explore other territories and dream about lost continents over the horizon, but also their attention has always been captured by the stars in the night sky. Were they places where dead souls would one day go? Were they faint and far away or bright and nearby? Were they places like Earth with inhabitants of their own? And how many of them were there?

Amongst the ancient experts there were conflicting opinions. The atomists like Lucretius and Epicurus believed that an infinite number of worlds of all sorts truly existed. This followed naturally from their belief in the existence of infinite numbers of atoms existing in an infinite Universe. There was no reason to suppose that any part of the Universe was different on the average to any other part, and so we would expect planets like the Earth and stars like the Sun to exist in infinite profusion. In the fourth century BC, Epicurus was spelling out this belief simply and clearly:

> 'There are infinite worlds both like and unlike this world of ours. For the atoms being infinite in number, as was already proved, are borne on far out into space. For those atoms which are of such nature that a world could be created by them or made by them, have not been used up either on one world or a limited number of worlds . . . So that there nowhere exists an obstacle to the infinite number of worlds.'[3]

As a counter to this view, there was the ancient authority of Aristotle who maintained that the Universe was composed of a finite amount of matter, with a definite centre and symmetry that was essential to the nature of things (Figure 9.1). An infinite Universe that contained many worlds like the Earth would lack the balance provided by a single centre. And so Aristotle rejected the idea that the Universe could be infinite, and with it the possibility of an infinite number of other worlds beyond the Earth.[4]

Over the next two thousand years Aristotle's ideas became the standard model for philosophy, theology, and science. They laid great emphasis upon the purpose of things as an essential part of any full explanation for their existence and form. The rival atomist doctrine of

Fig 9.1 *Aristotle (384–22 BC).*[5]

the chance movement of atoms as the basis for the explanation of everything was anathema to the Aristotelian picture and fell into abeyance until its revival in Europe in the early fifteenth century. It was natural that Aristotle's picture of purposeful change would become joined with the early Christian theology. Atomism never did. The chance movement of atoms was seen as an affront to the omnipotence of God, and the opposition of the early atomists to the relevance of the Greek and Roman gods and the idea of life after death allied it with more strident atheistic views.

As the synthesis of Aristotle's philosophy and Christian theology developed, a number of interesting problems emerged out of the issue of an infinity of worlds like the Earth. The first was the question of God's sovereignty over the Universe and the extent to which the created order ought to reflect the character of the Creator. There were those who took a literal approach to Holy Scripture, disbelieving what was not positively affirmed there; they argued that the creation stories in the book of Genesis made no mention of other worlds or other living creatures except for those on the Earth, and so they could not exist. God rested on the seventh day: He didn't create other universes. Others opposed this prescriptive negative theology to argue that while we should believe what Scripture affirms, we should not necessarily disbelieve what it does not. Thus, the lack of scriptural reference to the other planets in our solar system does not mean that they do not exist.

Some argued that it was acceptable to believe that God only created a finite Universe despite having the power to make an infinite one because He does not, and need not, do everything that it is possible for Him to do. Others insisted that the infinite power of God must be manifested in His works of creation and it was limiting to imagine that He would restrain His creative activity to a finite portion of a potential infinity.

The great German philosopher Immanuel Kant, before he became a critical philosopher of knowledge and how we obtain it, was an imaginative astronomical theorist. In 1755, when he was thirty-one, he proposed new schemes for the formation of our solar system and the spread of life and mind throughout the Universe. Kant believed that the Universe was infinite and contained an infinite number of inhabited worlds.[6] The second belief followed inevitably from the first because he could not conceive of a Universe as a 'vast unbounded desolate Negation of Beings'. Here is why he thinks that an infinite Deity would not be represented by a Universe that was merely finite:

> 'Now it would be absurd to represent the Deity as passing into action with an infinitely small part of His potency, and to think of his Infinite Power – the storehouse of a true immensity of natures and worlds – as inactive, and shut up eternally in a state of not being exercised. Is it not much more reasonable, or to say it better, is it not necessary to represent the system of creation as it must be in order to be a witness of that power which cannot be measured by

> any standard? . . . Eternity is not sufficient to embrace the manifestations of the Supreme Being, if it is not combined with the infinitude of space.'[7]

The argument was not really a new one, and the key point was not whether God could do infinite things but whether he chooses to. This type of argument was made most famously by St Augustine in his work *The City of God*, where he argues that those who believe that God must be active everywhere in the Universe will inevitably come to the conclusion that an infinite number of other worlds and beings exist. What worries him most about the idea that there are numberless other worlds elsewhere in space are the implications of the infinite replication paradox that we discussed in the last chapter. For the Christian doctrine of the Incarnation would require that it happened infinitely often, in different places, and that it was repeated infinitely often in time if one also subscribed to the Stoic view that the Universe ran through an unending series of cycles. This he finds an unacceptable consequence:

> '[are we to believe that] this same Plato, and the same school, and the same disciples existed, and so also are to be repeated during the countless cycles that are yet to be – far be it, I say, from us to believe this. For once Christ died for our sins; and rising from the dead, he dieth no more.'

This argument would cut no ice with the atomists of course, but it provides an interesting early example of how the status of one unique event in human history was regarded as sufficient to deny existence of an infinite number of worlds elsewhere. There were other arguments that Augustine could have used instead. He could have allowed the Universe to be infinite, and even to contain an infinite number of worlds, but argued that life only arose on the Earth and so no Incarnation was necessary, or possible, elsewhere. This deals with the world that is infinite in spatial extent, but it doesn't kill the objection to the eternal recurrence. For even if there is life only here on Earth, the history of the Earth will repeat infinitely often into the future (as it did infinitely into the past) and these copies would include the Incarnation.

Augustine's appeal to the uniqueness of the Incarnation as an argument against the possibility of other worlds can still be found in the sixteenth-century writings of Philip Melanchthon, who was a noted expositor of Martin Luther's reforming theology. Yet, in the meantime, the Catholic Church had experienced a more complex sequence of claims and counterclaims. Although Copernicus did not advocate an infinite Universe or a multitude of other Earths, others took this as a consequence of a Copernican system which removed the Earth from the natural Aristotelian centre of things. Once the centre is displaced, Aristotle's arguments become conspicuously weak, and they were gradually whittled away by critical commentators like Giordano Bruno, John Wilkins, and Henry Gore.

The French writer Michel de Montaigne[8] introduced a new dimension to the properties of many worlds by raising the question of why the laws of Nature need be the same in these other worlds. Why not allow them to explore all possible legislations?

> 'Now if there are many worlds, as Democritus, Epicurus, and almost all philosophy has taught, how do we know whether the principles and rules of this one apply similarly to the others? Perchance they have a different appearance and different laws.'

This type of argument might well undermine any conclusions drawn about what was possible in other worlds, or it might lead to the yet more speculative notion that all possible laws of Nature arise somewhere amongst the infinite collection of possible worlds. As we shall see, this is a familiar idea today, but it is handled in a rather different way. For we expect that amongst the panoply of possible worlds governed by all possible laws, there will only be some where complexity and life are possible. Only in one of these privileged worlds could we find ourselves. We must inevitably judge ourselves to be special in *some* sense and fortuitous that the laws in our world have fallen out so as to permit life to exist. While Copernicus's idea that our position in the Universe should not be special in *every* sense is sound, it is not true that it cannot be special in *any* sense.

Another new distinction in these arguments was introduced by the young English clergyman and scientist John Wilkins, whose instant book (he claimed to have written it in a week), published in 1638, when he was just twenty-four, under the title *Discovery of a World in the Moone,* was intended to show that Christian faith and reason were not opposed and that it was consistent with both to believe in an inhabited lunar world and life on other worlds throughout the Universe. Wilkins was careful to distinguish two sorts of 'worlds' that generally get run together in these discussions.[9] On the one hand he says the term World can be used to describe an entire Universe, containing all the stars and the Earth; but on the other hand it may be used more particularly to refer to a particular celestial body, like the Moon or another habitable planet.

The scientist who brought down the greatest retribution upon himself for his espousal of an infinity of worlds was the controversial figure of Giordano Bruno. Bruno believed firmly in the existence of an infinity of Earthlike worlds in the Universe. As we have seen, he took issue with Aristotle's picture of the Universe on all fronts. He maintained that Aristotle's division of the Universe into a finite sub-lunar realm of physical things and a celestial realm of ethereal things was a denial of the unity of the Universe. And then he picked to pieces all of Aristotle's arguments against an infinite Universe and a plurality of worlds. Aristotle had argued that with more than one world there would be an asymmetry – only one could be at the centre and only in this one-world Universe was ordered motion possible. But Bruno points out that in an infinite Universe there is no single centre and no single circumference and so Aristotle's arguments fail. In all these develop-ments we see the close link between the nature of an infinite Universe and the question of many inhabited worlds. Arguments for and against one idea were liable to be wielded against the other and there were few attempts to disentangle them. However, we must remember that the scientific conception of the Universe was a good deal smaller then than now. Stars and planets were the basic ingredients of the Universe and both were closely linked to the requirements for life.

OUT OF THIS WORLD

'The world is not enough'

James Bond movie title

The dispute over the plurality of worlds lives on. The matter is just as contentious now as it always was, but, fortunately, you don't get burnt at the stake for being on the wrong side of the argument any more. The two threads that Wilkins distinguished, involving separate universes and separate habitable worlds (or planets) in our Universe, lead to different arguments. We have a debate amongst biologists and astronomers of all sorts as to the evidence and likelihood of extra-terrestrial life. It is widely believed that simple life-forms – like bacteria – should be prevalent in the Universe, and may even still exist on nearby planets like Mars or the moons of planets like Saturn and Jupiter, but when it comes to the question of intelligent life, and life that is more advanced than ourselves, we have only speculation. We do not understand the processes by which life came to exist and evolve on Earth; we do not understand what consciousness is with sufficient precision to make any worthwhile predictions about the likelihood of it arising on other planets under different conditions. The problem is that we are trying to generalise from just one example. If life of any sort were found elsewhere in our solar system, it would set in train a wave of speculation about life elsewhere.

The other strand of the many-worlds argument is more speculative still. It asks whether there can exist a multiverse of other universes – not just other planets within our Universe. These other universes can have different laws and contents. We shall see that the distinction between the different varieties of other worlds is not quite so clear cut. There is scope within an infinite universe for things to be very different elsewhere in our Universe, even though, technically, there is just one Universe. What are the possibilities that modern cosmologists take seriously and why?

There are two motivations for thinking about other universes. The first, and most recent, is the possibility that they are a requirement of modern physics, that if one Universe exists, then it is required that others do also. The other imperative is quite different. It has been known for some time that our observable Universe possesses many tantalising life-supporting properties. If the expansion rate or level of uniformity of our Universe were very slightly changed, or if the constants of Nature had taken on slightly different values, or if the number of large dimensions of space had been other than three, then any type of atom-based complexity would have been impossible. Our Universe is a little like a cone balancing on its apex. Nudge its basic properties very slightly and everything goes wrong: no stars, no planets, no atoms, no complexity, no life. Consequently, we have come to refer to our Universe as 'fine tuned' for life.

What do we make of this? Some responses are shown in Figure 9.2. If there can be one and only one possible Universe, then we have been very lucky and there is nothing more to be said. Things are as they are and they allow life to exist and the evolutionary process on planet Earth has exploited that wonderful possibility. Thirty years ago, this would have been the favoured option. Physicists looked forward to a Theory of Everything which would show that the Universe could only be one way. It was expected to be like a jigsaw puzzle with a unique solution. Change the shape of a single piece and there is no way to fit the puzzle together. However, as time has gone on, and

Some options
Only one universe is logically possible – we were lucky that life could exist in it
All possible universes exist – we inevitably live in one of those where life is possible
'Life' is much easier to produce than we think and can exist in very different forms to our carbon-based chemical variety – almost any sufficiently complex universe can produce it
The constants and laws of nature are not fixed – there is an evolutionary process that inevitably results in the present fine-tuned state
The universe must eventually contain life. For some as yet unknown reason self reference is a necessary attribute of the universe
The universe is infinite and diverse in its properties – there always exist places where the coincidences needed for life to exist will arise.

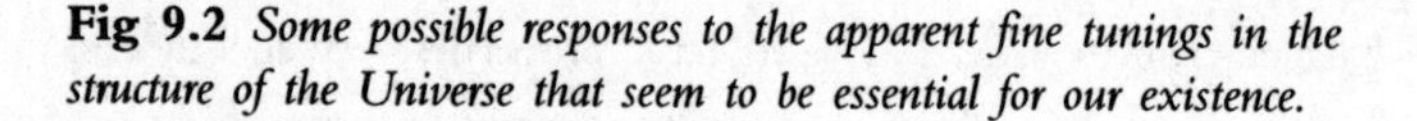

Fig 9.2 *Some possible responses to the apparent fine tunings in the structure of the Universe that seem to be essential for our existence.*

parts of the elusive Theory of Everything have started to show us some glimpses of its likely character, this simple expectation of an inflexible strait-jacket of a theory that fixes and explains everything has started to look less and less likely. Instead, a gap has opened up between what information is needed to specify the Universe – its laws, constants, and cosmological properties – and what can be contained within a viable Theory of Everything. As the first candidates for a Theory of Everything emerged from string theory in the 1980s, the hopes for a unique theory slipped away as it appeared that there were five candidates. Then things changed, and these five theories appeared to be different views of some deeper underlying 'M' theory (M stands for Mystery) that has yet to be found. Then, it became clear that there was an infinite continuum of these theories that interpolated between the five and which included the M theory as well. A huge landscape of possible Theories of Everything has opened up.[10]

Far more of the Universe's properties seem to have a random or flexi character, able to be other than what they are, without doing violence to the underlying Theory of Everything. Only a few of Nature's most fundamental constants and properties may be programmed-in by the special mathematical character of the Theory of Everything. The rest appear to arise at random from a vast range of possibilities. So far we don't know how to determine which are the 'most likely' combinations. Even if we could, it is not clear that that is a useful thing to know. If the most likely combination of properties describes a universe which does not allow life to exist, then it cannot describe our Universe (Figure 9.3). Life-supporting combinations of properties may be very rare in the sea of all possibilities, yet our Universe must be one of those rare fish in the sea.

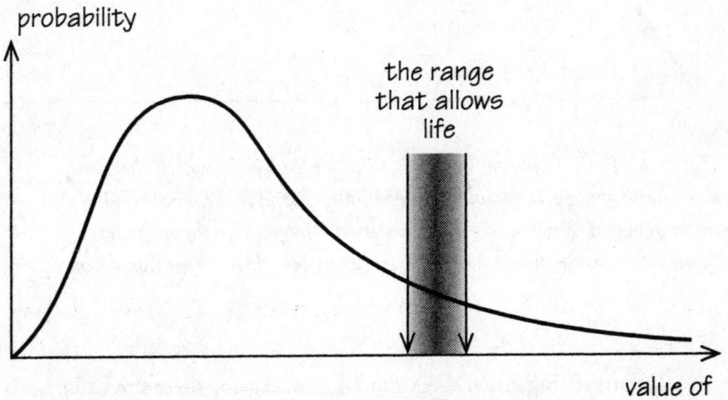

Fig 9.3 *A possible form for the probability that a constant of Nature takes different possible numerical values. The most probable value predicted by a theory may not fall in the narrow intervals that allow life to exist.*

Our prime candidate for a family of Theories of Everything predicts that the Universe has ten dimensions, not three. Only in ten dimensions do all the disruptive infinities and mismatches between the different forces of Nature disappear. The only way to reconcile this state of affairs with the fact that our Universe stubbornly persists in having only four dimensions – three of space and one of time – is to suppose that the 'extra' six dimensions are all space-like and have remained imperceptibly small. Only three have become astronomically large, as shown in Figure 9.4.

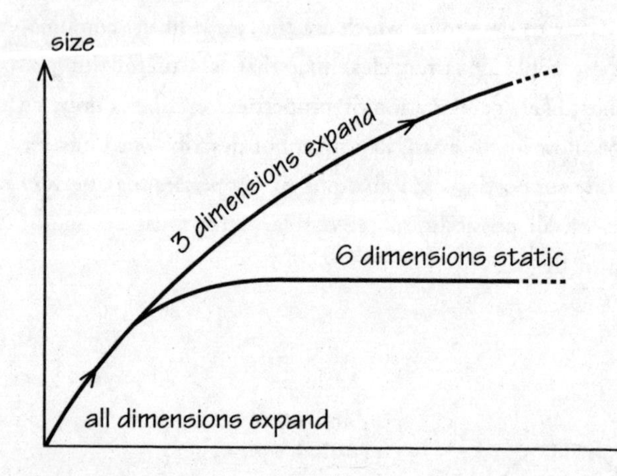

Fig 9.4 *If our Universe possesses nine dimensions of space then only three have become large and the remaining six must be small and unchanging in size to very high precision. Large or changing extra dimensions would have observable effects in our three-dimensional space.*

Why the 'big three'? We don't know. Again, there could be some deep reason why three, and only three, dimensions can grow large, or it may be simply a random outcome of events near the beginning of the Universe. If the latter, then there is no reason why the number should be the same everywhere. In some other parts of an infinite

Universe there may be three big dimensions of space, elsewhere two, elsewhere nine, and so on. In each region things will be different. All we can say is that we have learnt that atoms and life can only exist where three dimensions of space become big.[11] Other dimensionalities may well exist elsewhere in an infinite Universe, but those regions will be stillborn, devoid of observers.

We begin to see another approach to the life-supporting fine tunings emerging here. If the constants and other properties of the Universe can fall out in different ways in different places within an infinite universe, then we must necessarily find ourselves inhabiting one of those regions where things fell out in a life-supporting way. No matter how finely tuned things had to be in order to support living beings, we would have to see those fine tunings where we exist because we could observe no other sort of local universe.

There are very respectable theories of the very early history of the Universe which lead inevitably to an infinite Universe with markedly different properties from place to place. The most popular is the eternal inflationary universe theory of Alex Vilenkin and Andre Linde. It is a generalisation of the inflationary universe theory, which provides a good explanation for many of the large scale properties of our Universe. This theory proposes that during the first 10^{-35} second of the Universe's history, it experiences a brief period of acceleration as a result of the presence of new forms of matter, which are predicted to exist at very high temperatures. These novelties quickly decay into radiation and the Universe resumes its more lugubrious career of decelerating expansion (Figure 9.5).

The overall effect of this temporary surge in the expansion rate is to enable the whole part of the Universe that we see today to have emerged from the expansion of a tiny quantum fluctuation that was small enough to keep smooth, save only for the presence of tiny quantum statistical variations from place to place.[12] As the Universe's expansion surges ('inflates'), these characteristic variations in the temperature and density of the Universe become stretched over large regions of space.

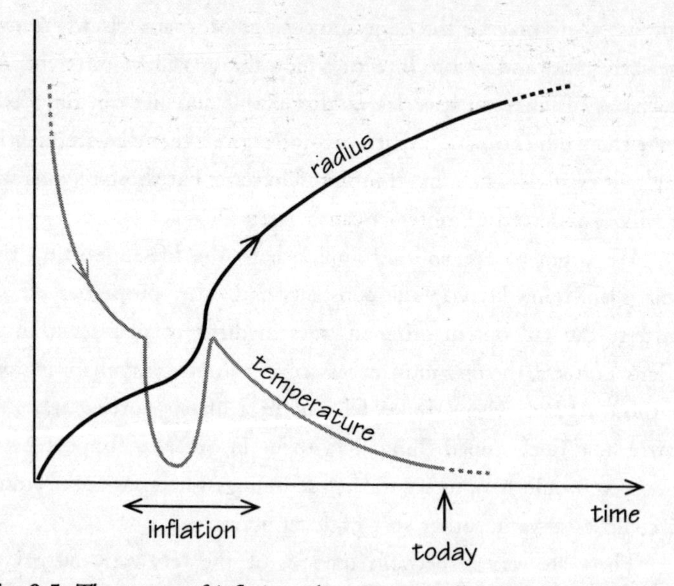

Fig 9.5 *The process of inflation takes a very small region of space and expands it in size by a large factor. In this way, tiny regions near the beginning of the Universe which were small enough to be keep very smooth by physical processes acting at the speed of light can easily grow to become larger than the scale of the entire visible Universe today. Here, we see the way in which the scale and the temperature of one of these regions vary with time. After cooling dramatically, the inflated region is heated up again by the decay of the unusual forms of matter and energy that sustained the brief interlude of accelerated inflation. It then resumes cooling until the present.*

They imprint their signatures on the visible radiation today in the form of small temperature differences from one direction to another on the sky. Assuming the laws of physics are the same in this environment as they are now, we can predict what these tiny fluctuations should be like and then go and see whether the expected variations exist in the radiation that is left over from the beginning of the Universe's expansion.

The detection of these little variations in the temperature of the radiation from the early stages of the Universe has been a major goal of observational cosmology for the last twenty years. It has become a remarkable success story. A series of high-tech experiments, flown in high-flying balloons and in satellites, has built up a consistent picture of what the fluctuations in the background radiation look like. The most dramatic of these observations came with the announcement in early 2003 of NASA's Wilkinson Microwave Anisotropy Probe (WMAP) observations.[13] These can be seen in Figure 9.6.

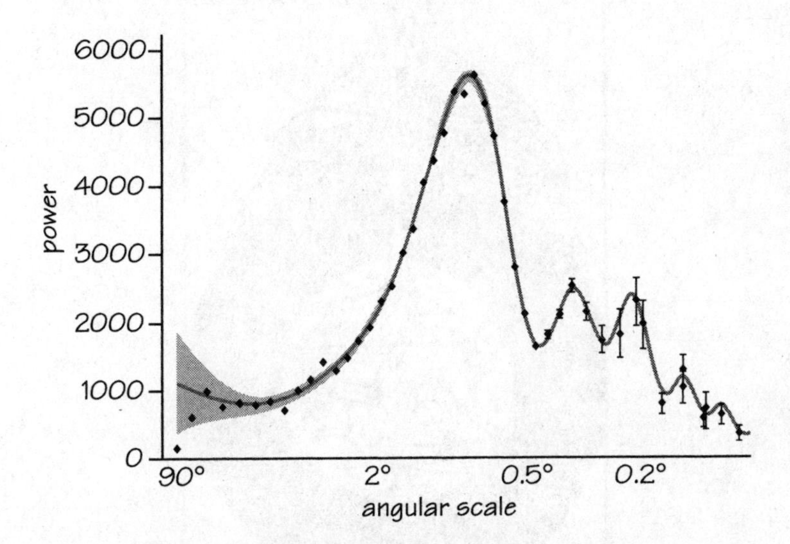

Fig 9.6 *WMAP satellite observations of temperature fluctuations in the background radiation versus scale compared to those predicted by the simplest theory of the inflationary theory which are shown by the solid line.*

This picture shows the remarkable and detailed agreement between the predictions of the simplest theory of inflation and the WMAP observations. The measurement uncertainties of the WMAP observations are very small and there is very little scope for the theory to fit the data with high accuracy by shifting the solid curve around. This provides excellent evidence that inflation is at least part of the truth about what happened soon after our Universe began to expand. It is especially striking that events in the first 10^{-35} second of the Universe's history produce clear signals that can be detected like fossils today. It would have been even more fantastic to the fifteenth-century author of the striking resonant image in Figure 9.7.

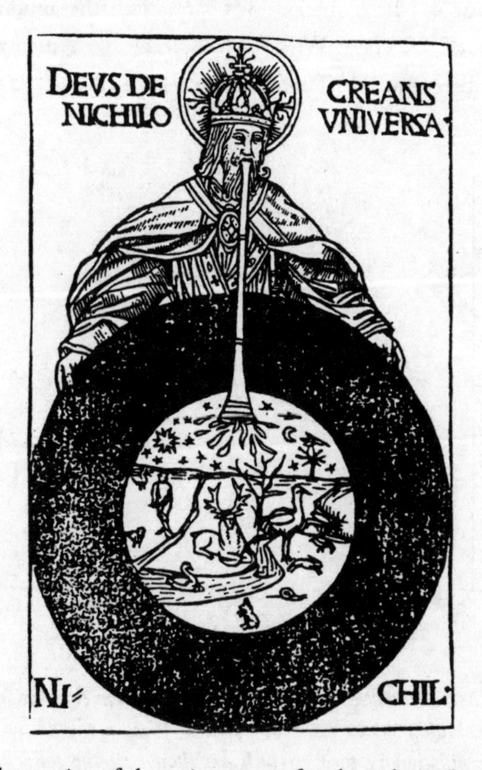

Fig 9.7 *The creation of the universe out of nothing depicted in* The Seventh Day of Creation *by the French philosopher Charles de Bouelles in 1510. God is breathing the breath of life into his embryonic universe – an early theory of 'inflation'?*

It is this impressive agreement between the simple predictions of the inflationary universe theory and such a large number of high-precision observations that provokes us to take seriously the wider consequences of the inflationary universe for infinite universes.

INFLATION – HERE, THERE, AND EVERYWHERE

'Can we actually "know" the universe? My God, it's hard enough finding your way around in Chinatown.'

Woody Allen[14]

The infinite inflationary universe introduces new sources of variation into the structure of the Universe. The tiny bubble that inflates to become larger than the region that we call the visible Universe today is just one of many – infinitely many. Each will undergo a different amount of inflationary expansion. The result will be an expanded Universe where the variations in structure from one small region to another will have become magnified by the expansion into differences in density, temperature, and expansion from one part of the Universe to the next. We appear to live in one of the regions that have inherited their 'genetic code' from a single bubble; but if we could travel far beyond the distance that light has had time to travel since the expansion began (more than 14 billion light years), we could expect eventually to run into another region that inherited a quite different structure from the inflation of a different bubble. An infinite inflationary universe will therefore have an alarmingly complex geography.

All the structural possibilities for the expansion, density and temperature of our observable Universe may be played out across the infinity of space that results from the chaotic inflation of an infinite universe. Somewhere amidst this never-ending foam of inflated bubbles there will be regions, like our own, where conditions fall out right for the development of life. Deviations from smoothness in the density of matter will be neither so strong that everything ends up prematurely forming black holes, nor so weak that no islands of material ever separate out from the expansion of the Universe and everything just keeps expanding into a formless future. In the 'Goldilocks' regions, where matter is neither too dense nor too sparse, observers will be possible. If observers are, like ourselves, made out of the chemistry of atoms heavier than simple hydrogen and helium gases, then lots of time is needed. The inflated bubbles must grow big enough and old enough to allow for ten billion years of slow nuclear reactions to occur inside the stars, where carbon, nitrogen and oxygen and all the elements of complex biochemistry are formed from hydrogen and helium.

Physicists have explored the possibilities for this type of geographically complex universe with its many islands of possibility realised within our single Universe. It has emerged that the different inflated bubbles can end up being different in many more ways than in their density and temperature. When the short bout of inflation that expands them comes to an end, it can leave the bubbles in very different states (Figure 9.8). The nature of those forces and constants in each region will end up being randomly chosen by the cooling down process. In some regions there will be only a force of gravity, in others there will also be the weak, strong, and electromagnetic forces that we see in our world; perhaps in others there will be additional strong forces that do not exist in our world or are imperceptibly weak. The laws of Nature will have given rise to a number of different territories, each governed by their own randomly chosen by-laws.

This structure is remarkable because, for the first time, it gives us a positive reason to expect that our Universe is not the same everywhere. Beyond our visible horizon we expect its structure, perhaps even its laws and fundamental constants, to vary in all possible ways. In an infinite universe they will explore the whole gamut of possibilities that are open to them. Our Universe's laws and constants appear like by-laws governing a district in a vast cosmopolis too diverse to imagine.

We have seen that the ancients had fantastic visions of such variations. They could not rule them out, except by introducing inflexible theological or philosophical dogma. But the inflationary Universe is

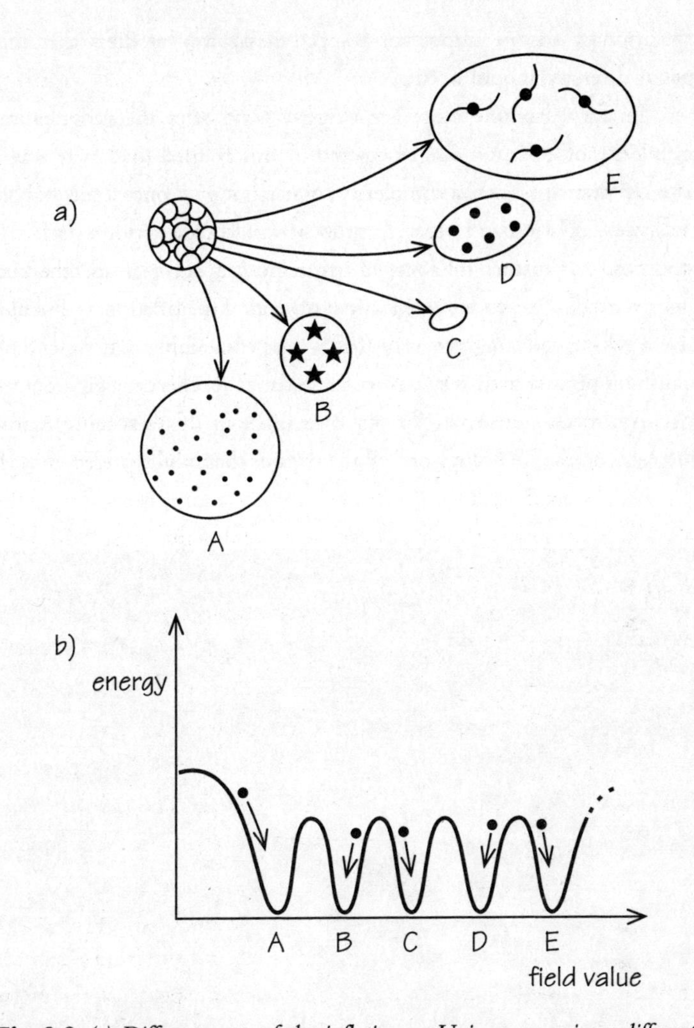

Fig 9.8 *(a) Different parts of the inflationary Universe experience different amounts of inflation. (b) Symmetries which govern some of the laws and constants of Nature can fall out differently according to which of the energy 'valleys' the state of the Universe comes to rest in. Each valley corresponds to a different form for the laws of physics that will emerge as the Universe expands and cools down further. Parts of the Universe which fall into the states A, B, C, D or E will exhibit different physics and only some of them will allow living complexity to form.*

revolutionary in one important respect. It *positively* predicts that this spatial diversity should occur.

Inflation has one more ace to play. Soon after the geographical complexity of inflation was discovered, it was realised that there was a yet more dramatic *historical* complexity to match it. For, once a tiny bubble underwent inflation and began to grow, it would create within itself the conditions for further inflation of its many tiny sub-regions to occur. Once started, this process would have no end. The inflationary bubbles have a self-reproducing property that is perpetuated like a never-ending branching process with very particular statistical properties (Figure 9.9).

Again, we can search for our own place in this cascading reproductive process. We occupy one of the regions that has expanded enough

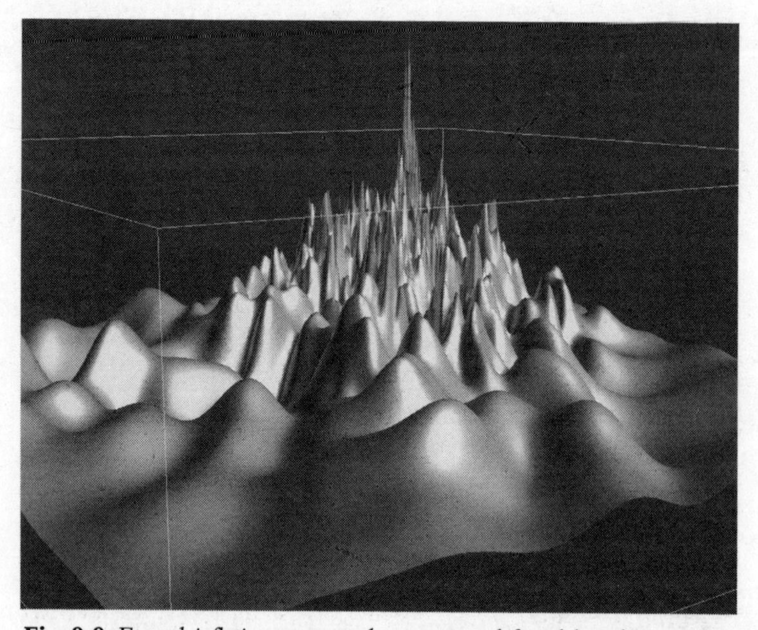

Fig 9.9 *Eternal inflation represented as an eternal fractal branching process – the 'Kandinsky universe'. The hills represent regions that inflate, with further hills spiking up from them in a never-ending fractal process that fills space.*[15]

for there to be time for stars to form and biochemical elements to arise. While the theory predicts that the process has no end, it remains to be proved whether or not it ever had a beginning. We expect that individual regions may have 'beginnings', but the process as a whole, and time itself, need not. This process is called 'eternal' inflation.

As in the chaotic inflationary process, each of the inflated bubbles may carry a different number of forces of Nature, different values for some (or even all) of the constants of Nature, and different numbers of dimensions of space and time. This is tantalisingly close to providing a sweep of the collection of all logically possible alternatives for the structure of universes within a very wide range of possible worlds.

These examples are important to our understanding of the idea of all possible worlds because they show how the laws of physics can produce a realisation of an infinite range of logically possible conditions within our Universe, if it is infinite in size, *without us having to appeal to metaphysical notions like 'other' universes existing in parallel realities.* One infinite universe contains enough room to contain all these possibilities. This is the conservative multiverse option.

Can this idea ever be tested by observation? At present it is hard to imagine how a direct observational test could ever be possible. The different cosmological structure, the different laws of physics and constants of Nature, all by definition occur beyond our visible horizon. We can never see them with our telescopes right now. Although we might in principle suddenly find these other worlds swimming into view in the far far future, that would involve the most staggering coincidence: why should our bubble have inflated by this minimal amount that gives *us* enough time to evolve and build telescopes?

It is overwhelmingly more likely that our bubble inflated by far more than the extent of our visible Universe today. We would never run into it; which is a good thing because, if we did, the consequences would be catastrophic for our Universe. However, despite this forecast, we might have some possibility of gaining some indirect evidence that other worlds, under different natural legislations, exist beyond our horizon. Suppose that a theory which allows these diverse other regions to exist requires each region to carry a certain signature which can be observed, whereas if they do not exist then the signature is absent. The signature might be a feature in the pattern of fluctuations in the radio waves left over from the early Universe which we can observe. In that case we could have a way of ruling out the existence of the different other regions if the underlying physical theory was supported by other evidence. This is probably the best that we can hope for. After all, there is no reason to believe that the Universe was constructed for our convenience. We have no special right to expect that all truths about the Universe can be tested by observations that are within our reach: that really would be an anti-Copernican outlook.

CONSCIOUS INTERVENTIONS – MEN IN BLACK

> 'In general the Star Maker, once he had ordained the basic principles of a cosmos and created its initial state, was content to watch the issue; but sometimes he chose to interfere, either by infringing the natural laws that he himself had ordained, or by introducing a new emergent formative principle, or by influencing the minds of the creatures by direct revelation. This according to my dream, was sometimes done to improve a cosmical design; but, more often, interference was included in his original plan.
>
> Olaf Stapledon[16]

If inflationary bubbles can create within themselves the conditions needed for their self-reproduction to occur, then could *we* engineer those conditions to occur at will? A frightening thought. It raises the question of what impact intelligence can have on the cosmic environment. It is clear that human beings, by virtue of their intelligence and ability to foresee some of the consequences of their actions, have had a huge impact on the terrestrial environment. We can shape the way parts of the Earth's surface develop, change the weather, alter the populations of other living things, and even change the course of natural evolution on near-by planets.

Scientists were always used to excluding themselves from the equations and experiments they were performing. The act of observing the world was akin to being a bird-watcher with a perfect hide. In the physics of the nineteenth century the act of observing the world had no effect on what was being observed. That was to change irreversibly with the discovery of the quantum theory. However, the question of what would happen if freewill and intelligence were to intervene in a physics experiment did not have to wait quite so long.

During the nineteenth century, the celebrated British physicist, James Clerk Maxwell, created a landmark thought-experiment. Imagine, he proposed, that a chamber is divided into two parts by a wall running across it which has a small trap-door in it, as shown in Figure 9.10. If the door is left open, then eventually the molecules will bounce off the walls of the chamber, and off each other, so often that an equilibrium will be created and the temperature (a measure of the average speed of the molecules) will be the same throughout the chamber.

Now, Maxwell asks, what will happen if a little 'demon' exists inside the chamber who is able to detect fast- and slow-moving molecules. He opens the trap-door so as to let only fast-moving molecules into one end of the chamber and slow-moving molecules into the other end. What will happen? Instead of an equilibrium being created, one end of the chamber will fill up with fast molecules, and so get hot, while the other end will get cold! If we chose to remove the door after the temperature difference had been created, we could use it to drive a machine, so getting work done for nothing! A funda-

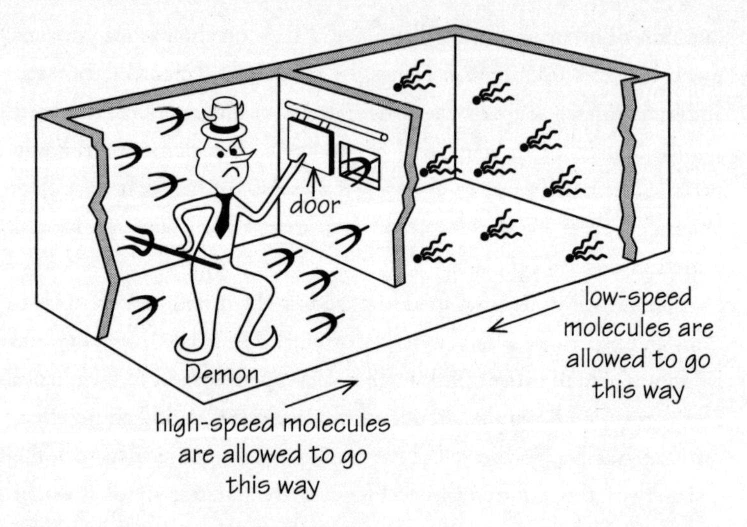

Fig 9.10 *James Clerk Maxwell's sorting demon supposedly separates fast and slow-moving molecules by opening and shutting a trap-door in the dividing wall of the chamber. If all the fast molecules end up at one end then it will become hotter than the other end, seemingly in defiance of the laws of thermodynamics.*

mental law of thermodynamics appears to have been violated.

Fortunately, when this paradox is examined closely, it turns out that no violation of the laws of thermodynamics is possible when all the energy requirements of the demon are included in the story. He has to do work to detect the motions of the molecules, open and close the trap-door, and then wipe his slate of information clean again ready to make the next measurement. But, interesting as the resolution of this paradox is,[17] our main interest in it here is simply as an example of the sort of problem that can arise when conscious agents are allowed to appear in scientific 'experiments'. The arrival of freewill introduces a new ingredient that must be counted into the equations in some way. It is not something extra-scientific. It just requires careful thought to get it right.

The first consideration of direct conscious interventions in cosmology was made by the University of Arizona cosmologist, Edward

Harrison. He took his cue from the fact that there had been interest in the question of whether it might be possible to 'create' a universe in the laboratory. Cosmologists considered the conditions – rather extreme – that might be needed to make the next bout of eternal inflation happen here and now, to order. This is not as cataclysmic as it might sound because the 'created' inflationary bubble can just expand away at the speed of light and you might not notice anything. Although a simple recipe for doing this was not found, the difficulties encountered might, one day, be overcome. Let us suppose, Harrison suggests,[18] that these practical difficulties can indeed be overcome, and that civilisations vastly more advanced than ourselves are able to steer the course of eternal inflation in their vicinity by initiating the production of mini-universes in their laboratories.[19] And if they can do this they can guide the course of what would otherwise be random outcomes at the end of the inflation phase of the mini-universe's early expansion. This means that some (maybe all) of its resulting constants and forces of Nature can be selected as well.

Harrison speculates that if an advanced civilisation had this cosmological power, then it would try to make its baby universes more bio-friendly. They will have noticed, like us, that certain values of the constants of Nature and the structure of the Universe make it more likely that life will evolve and persist. Surely our universe-makers would respond by tuning these life-supporting coincidences to hold with greater accuracy in their new baby universes. They would maximise the chances of life developing in their futures.

What will happen in the long run? The more finely tuned baby universes will give rise to their own yet-more-advanced civilisations (perhaps inheriting information from their makers to accelerate them on their way). If they act like their predecessors, they will make more baby universes of their own and tune the relations between their adjustable constants and properties to be yet more bio-friendly than before. The result, Harrison, suggests, of many generations of this forced breeding[20] would be mini-universes in which the inhabitants found extremely fine-tuned apparent coincidences between the values of the constants of Nature and the structure of the universe which, if changed very slightly, would make life less likely or simply impossible. And that is rather like our Universe appears to be.

The far future of the Universe is another time and place where advanced technology can influence the evolution of increasingly larger parts of the Universe. We have already become familiar with the need for humans to develop a defence mechanism to deflect incoming asteroids and comets from hitting the Earth. Any such impact or very close encounter would be catastrophic for higher forms of life, and the climatic changes it created would change the entire course of the Earth's evolution. But if we look to the far future of the Universe, after all the stars and planets have faded, it is a long-standing challenge for cosmologists to find a way in which information-processing – a necessary feature of any time of intelligence – can continue indefinitely.[21] Again, one can envisage scenarios in which advanced civilisations can continue to extract energy from small differences in the expansion rate of the Universe from one direction to another. In effect this is cosmic tidal energy.[22] By releasing beams of radiation in different directions, they could be made to cool at different rates. The difference in temperature between the different beams would create a temperature gradient which could be used to drive a computing machine if the 'hardware' existed, perhaps in the form of elementary particles. If the Universe does not carry on accelerating forever, then this device enables an infinite number of bits of information to be processed in the infinite future. Such a computer would 'live' forever.

These examples show that when intelligent agents exert their influence on the behaviour of the Universe, or parts of it, sequences of events which would be judged highly improbable can suddenly become certain, if the will and the technological prowess exist. Whether they will are unpredictable political and sociological questions as much as scientific ones. We can only say whether certain events are possible. We cannot tell whether they will happen or not.

SIMULATED UNIVERSES

'All the world's a stage,
And all the men and women merely players'

William Shakespeare, *As You Like It*

Once conscious interventions are considered, there is an entirely new dimension to the multiverses problem. Recall that one motivation for considering the presence of other universes has been to understand why our visible Universe displays so many life-supporting 'coincidences' between the values of its numerical constants,[23] and so many advantageous contingencies that have fallen out apparently. Yet, contemplating all possible universes opens a bigger Pandora's Box. For amongst *all* possibilities there must exist universes populated by advanced beings who can create their own universes by virtual simulation. They would have computer power that differed from ours by a vast factor. Instead of merely simulating the formation of galaxies in their computers (as we can) they would be able to go further and watch the appearance of stars and planetary systems. Then, having coupled the rules of biochemistry into their astronomical simulations they would be able to watch as the evolution of life and consciousness ensued (all speeded up to occur on whatever timescale was convenient for them). Just as we watch the lifecycles of fruit flies, they would be able to follow the evolution of intelligent life, watch civilisations grow and communicate with each other, argue about whether there existed a Great Programmer in the Sky who created their universe and who could intervene at will in defiance of the laws of Nature they habitually observed.

Why would our advanced descendants do this? There are many motivations, all individually reasonable, which, when taken as a whole, become rather compelling. They will undoubtedly be at least as intellectually curious as we are. If they *can* do it, and like to argue about it, then you can be sure that eventually someone *will* do it. But they will also have historiographic reasons for simulating alternative realities. They will want to know what did happen, what might have happened, and what could not have happened, in the past if the gaps in their knowledge of it were filled in different possible ways.

Then again, simulated realities might just be a branch of their entertainment industry. After all, the most demanding programs and graphics packages run on our computers are not educational programs, business spreadsheets, or mathematical equation solvers; they are state-of-the-art computer games. That is what tempts investors in the industry.

The most alarming thought about such a simple future scenario is that *we* might be living in someone else's simulation right now. But then again, this is not such an outlandish idea as it might at first appear. Is it not very similar to many religious beliefs in which God is the Great Programmer who can choose to intervene in the world occasionally after setting it going (as in orthodox Christian doctrine), or choose not to intervene after the start (as in Deism)? Nor is this scenario unlikely. Once a single advanced civilisation is capable of creating simulated realities that are complex enough to contain observers, then an infinite number of different ones becomes possible. Thus a conscious observer chosen at random[24] would be most likely to be an inhabitant of one of these second generation simulated realities.

The multiverse scenario is favoured by many cosmologists as a way to avoid the conclusion that the Universe was specially designed for life by a Grand Designer. Others see it as a way of avoiding having to say anything about the problem of fine tuning at all. But now we see, once conscious observers are allowed to intervene in the Universe, rather than being merely lumped into the category of 'observers' who do nothing, that there is a new problem. We end up with a scenario in which the gods reappear[25] in unlimited numbers in the guise of the simulators who have power of life and death over the simulated realities that they bring into being. The simulators determine the laws that govern their worlds. They can change the laws; they can pull the plug on the simulation at any moment; intervene or distance themselves from their simulation; watch as the simulated creatures argue about whether there is a god who controls or intervenes; work miracles; or stealthily impose their ethical principles upon the simulated reality. All the time they can avoid having even a twinge of conscience about hurting anyone, because their toy reality isn't 'real', is it? They can even watch their simulated realities mature to a level that allows them to simulate their own realities.

There is a curious consequence of this course of events. Suppose the simulators, or at least the first generations of them, while having very advanced knowledge of the laws of Nature, still have incomplete knowledge of those laws. They may know a lot about the physics and computer programming needed to simulate a universe, but there will remain gaps in their knowledge or, worse still, incorrect deductions about the laws of Nature. These gaps and mistakes would of course be subtle and far from obvious, otherwise our 'advanced civilisation' wouldn't be advanced. But these flaws need not prevent simulations being created and running for long periods of time without serious problems emerging.

Eventually the little flaws will begin to take effect. Logical contradictions arise now and again; the laws in the simulations will appear to break down in small ways now and again. The inhabitants of the simulation are puzzled. They don't believe some of the observations that their simulated astronomers make which show their constants of Nature are slowly changing over time.[26]

Sudden glitches occur in their laws every so often in some of the simulated realities. Maybe this is because some of the simulators use a technique of simulation that they have found effective in all other simulations of complex systems: the use of error-correcting codes. If our genetic code were left to its own devices we would not last very long. Errors would accumulate: death and mutation would quickly follow. We are protected from this by the existence of a mechanism for error correction that identifies and corrects mistakes in genetic coding. Many complex computer systems possess the same type of internal immune system to guard against error accumulation.

If the simulators used error-correcting computer codes to guard against the fallibility of their simulations, then every so often a correction would take place in the state or the laws governing the simulation. Mysterious changes would occur that would appear to be governed by a different set of rules or by none at all.

But what will happen to the simulated realities set in virtual motion by simulators with only a partial knowledge of the laws of Nature needed to sustain them over long periods of time? Eventually, these realities would cease to compute. They would fall victim of the incompetence of their creators. Errors would accumulate. Prediction would break down. Their world would become irrational. They would be heading for the virtual analogue of death in a biological organism when errors accumulate to a lethal level. The only escape for them is if their creators intervene to patch up the problems one by one as they arise. Just like a computer system manager who emails patches to protect users against viruses, so the creators of a simulation could offer this type of temporary protection, perhaps by updating the laws of Nature as they understand them so as to include extra things they have learnt since the simulation was initiated. All this is very familiar to the owner of a home computer who receives, almost on a daily basis, updates and revisions to their operating system to protect it against new forms of invasion or fill gaps in its logic that its creators had not noticed.

Another potential problem for simulated realities is the temptation for simulators to avoid the complexity of using a consistent set of laws of Nature in their worlds when they can fake it so much more easily. When the Disney Corporation makes a film that features the reflection of light from the surface of a lake, it does not use the laws of quantum electrodynamics and optics to compute the light scattering. That would require a stupendous amount of computing power and detail. Instead, the simulation of real light scattering is replaced by plausible rules that are much briefer than the real thing but give a *realistic* looking result. We do that sort of thing all over the computer entertainment business and it is very likely that it would be the way that simulated realities would begin – indeed, to some extent they already have. There would be an economic and practical imperative for simulated realities to stay that way if they were purely for entertainment, and they would therefore be quite distinguishable from the real world – so long as we knew which was which to start off with.[27]

We might also expect that simulated realities would possess a similar level of maximum computational complexity across the board. The simulated creatures should have a similar complexity to the most complex simulated non-living structures – something that Stephen Wolfram (for quite different reasons, and nothing to do with simulated realities) has called the Principle of Computational Equivalence.[28]

One of the most common worries about distinguishing a simulated reality from a true one from the inside is the suggestion that the simulators would be able to take into account some difference one might think of ahead of time and pre-adjust the simulation to avoid the mismatch. This new simulated reality might then develop its own disparities with true reality but they would be plugged by another act of predestination. The question is whether this is possible in the limit. The problem is similar to that first considered by Karl Popper[29] to identify the self-referential limits of computers. The same argument was used in a different context by the late Donald MacKay, in many publications,[30] as an argument against the possibility of predestination which is knowable by those whose futures are being predicted. It is only possible for me to have a correct prediction of your future actions if it is not made known to you.[31] Once it is made known to you it is always possible for you to falsify it. Thus it is not possible for there to be an unconditionally binding prediction of your future actions. Clearly the same argument applies to predicting elections:[32] there cannot be a public prediction of the outcome of an election that unconditionally takes into account the effect of the prediction itself on the electorate. This type of uncertainty is irreducible in principle. If the prediction is not made public it could be one-hundred per cent correct.

So we suggest that if we live in a simulated reality we should expect occasional sudden glitches, small drifts in the supposed constants and laws of Nature over time,[33] and a dawning realisation that the flaws of Nature are as important as the laws of Nature for our understanding of true reality.

HOW SHOULD WE THEN LIVE?

'If you might be living in a simulation then all else equal you should care less about others, live more for today, make your world look more likely to become rich, expect to and try more to participate in pivotal events, be more entertaining and praiseworthy, and keep the famous people around you happier and more interested in you.'

Robin Hanson[34]

Unusual consequences seem to follow if we take seriously the idea that there exists an infinite number of possible worlds which fill out all possibilities. We can imagine how an extension of some of the science and technology we have at the moment would enable our successors to do some of these things. The implications for the nature of the world that we experience and its likely fallibility are striking, worrying even, and they take us back to the words the philosopher David Hume wrote at the end of the eighteenth century.

Hume's sceptical dialogues about many of the arguments for the existence of God that were fashionable at the time pick on presumptions in these arguments about the perfect nature of creation, the uniquencess of the Deity and so forth. Here is what he had to say about many worlds and their likely defects:

> 'You must acknowledge, that it is impossible for us to tell, from our limited views, whether this system contains any great faults, or deserves any considerable praise, if compared to other possible, and even real systems. . . . If we survey a ship, what an exalted idea must we form of the ingenuity of the carpenter, who framed so complicated, useful and beautiful a machine? And what surprise must we entertain, when we find him a stupid mechanic, who imitated others, and copies an art, which, through a long succession of ages, after multiplied trials, mistakes, corrections, deliberations, and controversies, had been gradually improving? . . . Many worlds might have been botched and bungled, throughout an eternity, when this system was struck out: Much labour lost: Many fruitless trails made: And a slow, but continued improvement carried on during infinite ages in the art of world-making . . .
>
> This world, for aught he knows, is faulty and imperfect, compared to a superior standard, and was only the first rude essay of some infant deity who afterwards abandoned it, ashamed of his lame performance; it is the work of some dependent, inferior deity; and is the object of derision to his superiors; it is the product of old age and dotage in some superannuated deity and ever since his death has run on at adventures, from the first impulse and active force which it received from him.'[35]

Hume's tongue-in-cheek scenarios conjured up images of a host of gods of varying degrees of competence creating universes of different quality, like apprentices attempting to copy the master. But if we replace his inferior and superannuated deities by simulators, then what he envisions is a realm where simulated universes abound: some good, some promising, others defective.

So, if all possible worlds exist and we are living in a simulation whose laws are not quite consistent with one another, does this make any difference? Indeed, should it make any difference?[36] It will be rather de-motivating if you are a (simulated) scientist trying to understand the way the world works. Anything could happen without reason. Not surprisingly, simulated realities are not welcomed into the scientific world-view. Philosophers take them more seriously and some have even tried to use them as arenas to discuss ethics. The problems they spawn are unusual. Robin Hanson has suggested the possibility of being in a simulated reality might produce its own influences on how you should act.[37] Simulated experiences, no matter how real they may seem, are much more likely to be brought to a sudden and unpredictable end than typical real experiences. This suggests to Hanson that 'all else being equal you should care less about the future of yourself and of humanity, and live more for today'. We are familiar with the fact that in films and the theatre the star is surrounded by other good actors who have to interact with the star, but as you move further away from the star then extras and low-paid jobbing actors can fill in the crowd scenes and non-speaking parts at low cost. Likewise in a simulated reality, the characters far from your action may just be fake simulated characters and you shouldn't worry too much about them. Above all, Hanson suggests, if you are part of somebody's simulation, be entertaining! Be famous! Be a pivotal person! This will increase the chances of your simulated existence continuing, and that others will want to resimulate you in the future. Fail to have these characteristics and you could become like the soap-opera character who quickly gets written out of the show, taking a long holiday in Vladivostok, never to return.

As we look around at the way people in the news do behave, we are drawn to the conclusion that we must be living in a simulation! However, none of this is very persuasive. How you should behave depends entirely on the moral stance of the simulators. If they like to be entertained then you will do well to be entertaining. But if they are dedicated to a noble purpose, you might have the greatest chance of continuing re-creation and simulation by being a martyr for a just and good cause. While we do not suggest that these codes of behaviour are taken seriously as the basis for how to live your life, they do bring sharply into focus the central problems of moral philosophy and our responses to them. If simulated realities are the commonest and we are in one of them, it would be worrying if they are simulations of the sort that we know. But why should they be? If we had always used the word 'simulation' to describe the result of a one-off act of creation by God then we are in a very similar situation, albeit with a Simulator of a greater sort.

These consequences for life in simulated realities have led some to regard them as strong arguments against the existence of other worlds. If most of these worlds are virtual, then they can display illusory laws of physics and we are on a slippery slope to knowing nothing at all because there is no reliable knowledge to be had. It is the counterpoint to solipsism and has many of the same paralysing consequences for any future thinking. If all possibilities are infinite and actual, then reality contains rather more than we can bear.

chapter ten

Making Infinity Machines

'To complete any journey you must complete an infinite number of journeys.'

James F. Thomson[1]

SUPER-TASKS

'How, if we were suddenly enabled to perform infinitely many tasks in a finite time, could we know that we were?'

Crispin Wright[2]

The speed of computers has been increasing steadily for decades following a simple trend that shows that speed doubles roughly every 24 months and the amount of processor power that one dollar will buy doubles in the same time. The fastest computer on Earth – the NEC Earth Simulator – can carry out 40 trillion calculations per second when it acts alone. Join several computers together and their speed grows in proportion to the number of linked machines. These speeds are almost unimaginably fast to you and me, but they are undeniably finite. The big question is not just whether improvement in computer speed will keep tracking the trend in Figure 10.1, first drawn by Intel's Gordon Moore, but whether a computer could ever carry out an *infinite* number of calculations in a finite time. Indeed, why should it be a 'computer' – could any machine do an *infinite* number of things in a finite time? Can there ever be an 'Infinity Machine' of any sort?

This idea sounds as if it has beamed down straight out of *Star Trek,* but philosophers and physicists have taken a surprisingly close interest in this question. The search for an answer has created a terminology all of its own, and one of these hypothetical tasks that involves the accomplishment of an infinite number of things in a finite time has been dubbed a *super-task.*[3]

Motivated by Zeno's great paradoxes of the infinite, the first modern scientist to address the problem was Hermann Weyl. Weyl was a multi-talented mathematician, physicist, and philosopher of science, who made fundamental contributions to just about every subject he touched. Educated in Germany, he ended his career in America as one of Einstein's distinguished colleagues at the Institute for Advanced Study in Princeton. Unusually for a mathematician, though, he was a finitist: he did not believe in the existence of actual infinities, even in mathematics. Thus, he had great sympathy for the revolutionary programme of Luitzen Brouwer who sought to outlaw the use of infinities in mathematics. Indeed, it was this sympathy for Brouwer's brand of finitism that ended Weyl's close friendship with David Hilbert.

Challenged by Zeno's famous paradoxes of motion, Weyl considered the infinite series of terms in which each was half the size of the previous:

$$\tfrac{1}{2} + \tfrac{1}{4} + \tfrac{1}{8} + \tfrac{1}{16} + \tfrac{1}{32} + \ldots$$

And so on forever. Now the sum of this series is equal[4] to 1. But, says Weyl, if a length of one metre really consists of the sum of the pieces

Fig 10.1 *Moore's Law of progress in computer technology has been a good approximation until quite recently.*

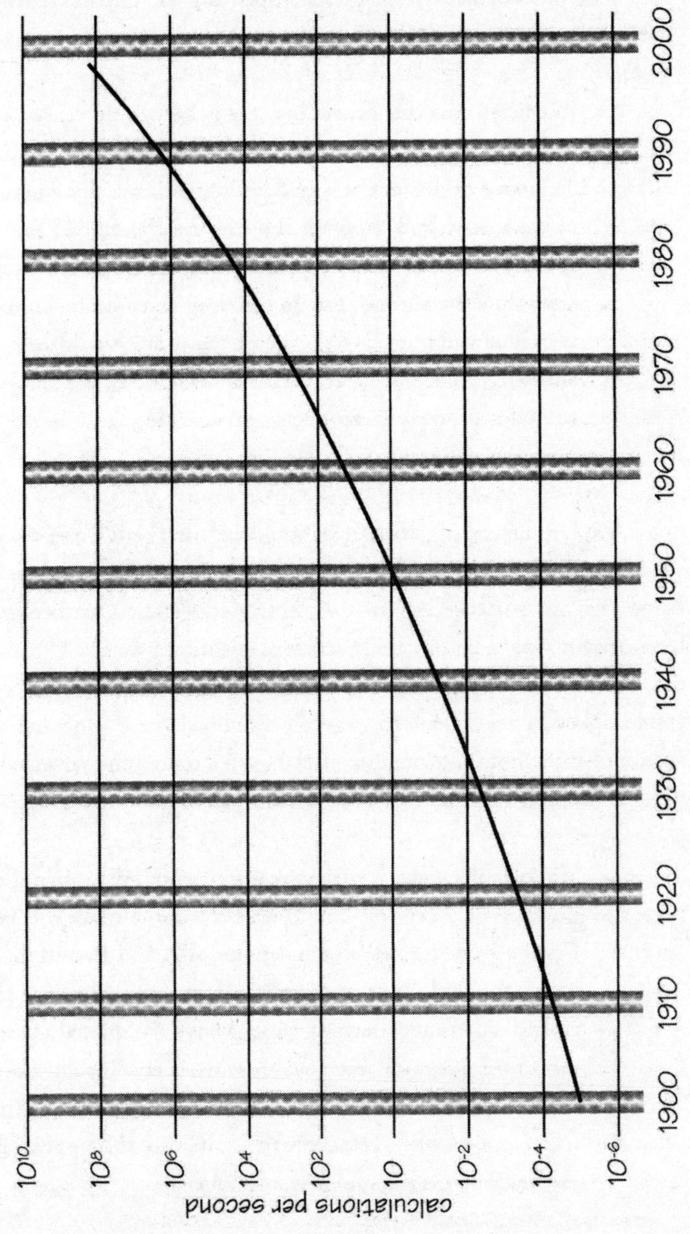
calculations per second
10^10
10^8
10^6
10^4
10^2
10
10^-2
10^-4
10^-6
1900
1910
1920
1930
1940
1950
1960
1970
1980
1990
2000

$1,000 of computing buys

$\frac{1}{2}$, $\frac{1}{4}$, $\frac{1}{8}$, . . . etc metre in length, chopped off the one-metre length, then we have a completable infinity and

> 'If one admits this possibility, then there is no reason why a machine should not be capable of completing an infinite sequence of distinct acts of decision within a finite amount of time; say, by supplying the first result after $\frac{1}{2}$ minute, the second after another $\frac{1}{4}$ minute, the third $\frac{1}{8}$ minute later than the second, etc. In this way it would be possible, provided the receptive power of the brain would function similarly, to achieve a traversal of all natural numbers and thereby a sure yes-no decision regarding any existential question about natural numbers.'[5]

Weyl didn't believe that there could be such a machine because he rejected the existence of actual infinities. But he didn't attempt to show that its existence led to some horrible logical contradiction, or even that it would be physically impossible to implement.

Five years later, Weyl's problem was revisited by the English philosopher James Thomson, who first dubbed these processes that accomplished an infinite number of things in a finite time 'super-tasks'. Thomson played the part of a modern Zeno, arguing that

> 'To complete any journey you must complete an infinite number of journeys. For to arrive from A to B you must first go from A to A', the mid-point of A and B, and thence to A'', the mid-point of A' and B, and so on. But it is logically absurd that someone should have completed all of an infinite number of journeys, just as it is logically absurd that someone should have completed all of an infinite number of tasks. Therefore it is absurd to suppose that anyone has ever completed any journey.'

The puzzle that Zeno first detected was that if we believe in a smooth continuum of time or space then any interval of time or space can be subdivided into an infinite number of pieces.[6] If you want to walk across the room, you must first walk half the distance, and then half of that distance, and so on, without end.

Thomson is trying to convince us that in order to complete a journey you must do something that is impossible and so you cannot complete the journey. The final conclusion seems impossible to believe, so there must be something wrong with the assumptions used to obtain it. Some people will say that it's incorrect to say that you can complete an infinite number of sub-journeys, but you don't need to.[7] Others have said that it's correct to say that you need to complete that infinite number of sub-journeys and, what is more, you can!

However, we can appreciate that in practice we don't move from place to place in this way. The infinite number of half-way houses are not physically distinct steps. This is not a case where an infinite number of things are 'done' in a finite period of time. For 'doing' things in the real world involves work and so, by the second law of thermodynamics, entropy production. This does not happen in the process of passing an infinite number of points along the way between two places. Suppose you were to cycle from Oxford to Cambridge. The distance is finite. It will take you a finite time to cover the distance. You will probably pass a number of mile-posts along the way, telling you how far you have to go. Their number is also finite. But suppose someone came along and increased the number of mile-posts, eventually making them infinite in number. You would still need to pass them all, but they would have no interaction with you. Their presence would make no difference to your cycling speed or the time and distance that you needed to complete in the saddle. They are innocuous infinities. By contrast, suppose there are stiles on the cycle path that need to be surmounted with your bike. The more there are on the route, the more work that must be done in overcoming them. An infinite number of such obstacles along a finite distance would not be an innocuous infinity. It would require an infinite amount of work to be done to overcome them and an infinite amount of entropy would be generated in the process.[8] Past discussions of paradoxes like Thomson's journey have not considered this property of his journey.[9]

Realistic super-tasks need actions to take place, rather than mere subdivision of the records by book-keepers. What sort of candidates have been suggested? Weyl didn't really suggest a blueprint. He merely took an innocuous infinity and proposed that there should be a machine that acted on cue at an infinite list of moments obtained by halving the interval of time since its last action. The real question is whether such an action is physically possible. In an attempt to put some flesh on the bare bones of Weyl's idea, Thomson then came up with a more specific device that has become known as the Thomson Lamp – although just giving it this rather official-sounding name doesn't mean it really exists![10]

In looking at such examples we must be careful to recognise that although a particular task, like taking a step, can be done without limit, this does not mean that the task of completing an infinite number of steps can be accomplished. Infinities are not just big numbers. They are qualitatively different to finite numbers, no matter how large those numbers may be. We have seen this distinction very graphically in the scenario of the Infinite Hotel, which can always accommodate new guests, even though it is full. Likewise in an infinite limiting process, the limit can have a property that is not shared by any of the individual items that add up to attain the limit.

RUBBING THOMSON'S LAMP

> 'On two occasions I have been asked, by members of Parliament, "Pray, Mr. Babbage, if you put into the machine wrong figures, will the right answers come out?" I am not able rightly to apprehend the kind of confusion of ideas that could provoke such a question.'
>
> Charles Babbage

Suppose that you have a reading lamp with a push-button that switches the light and off. If the light is 'off' to start with then if you press the button once, or any odd number of times, the lamp will be 'on'. Press it an even number of times and the lamp will be 'off'.

A little demon now appears and decides that he will press the button continually so as to leave the lamp 'on' for 1/2 a minute, then 'off' for 1/4 of a minute, 'on' for 1/8 of a minute, 'off' for 1/16 of a minute and so on. He will have pressed the button an infinite number of times after 1 minute[11] (see Figure 10.2). So the big question is: *Will the light be 'on' or 'off' after one minute?*

Your immediate reaction to this question is probably to claim that there is no such demon and no such lamp. It is physically impossible for such a sequence of switchings to take place. This is the answer of the physicist or the engineer. We know that quantum mechanics does not allow us to measure energies and time intervals simultaneously with arbitrary accuracy.[12] We would eventually be unable to measure the time interval before the next press of the button. And even if we could, we would not be able to respond at the ever-increasing speeds needed to press the button an infinite number of times in one minute.

Despite these replies, which no one really disputes, philosophers are still interested to know if there is a purely logical fallacy that would prevent an infinite number of actions being accomplished in a finite time. There is something odd about the question. It is tantamount to asking for the identity of the last member in a infinite sequence. Take for example the positive whole numbers: 1, 2, 3, 4, 5, 6, . . . and so on, without end. What is the largest whole number? Is it even or odd? This is the same question as the one that asks whether the lamp is 'off' or 'on' after one minute.[13]

Fantastic devices of this sort hold out an elusive crock of gold to the mathematician. There are so called 'irrational' numbers, like 'pi',

$$\pi = 3.14159 \ldots$$

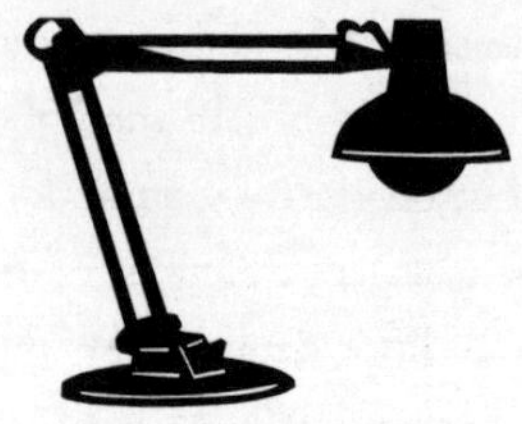

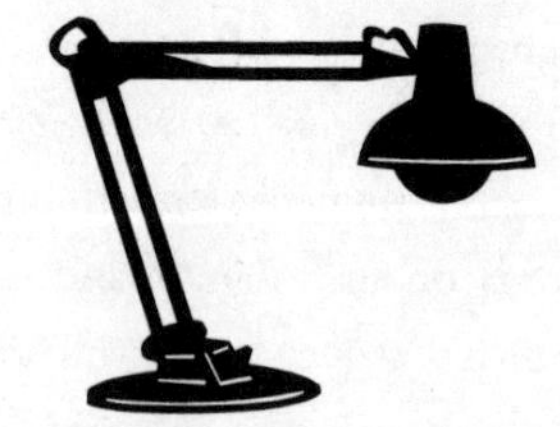

Fig 10.2 *Thomson's Lamp. Is it 'on' or 'off' after one minute?*

whose decimal expansion goes on forever. They cannot be written as the ratio of two whole numbers, no matter how large they are both chosen to be. We have arithmetic processes which can determine the decimal expansion of π to as many places as we choose.

There has always been a desire to see if there is some special property of this never-ending expansion, but alas none has been found.

Statistically it is a typical irrational number.[14] The late Carl Sagan framed a science fiction story, *Contact,* around the idea that there is a hidden message buried far downstream[15] in the decimal expansion of π. Only advanced, long-lived technological civilisations will have the computer power to reach and decode that message, which will fit them for the next plane of sentient existence.

Yet, a pi-in-the-sky infinity machine that was able to complete a super-task could determine the entire infinite decimal expansion of π in finite time. How? Just follow the work plan of the Thomson Lamp (see Figure 10.3). Print the first digit of the decimal expansion after ½ minute, the second after ¼ minute, and so on. An infinite number of printings would have been made after one minute has elapsed.

π = 3.14159 26535 89793 23846 26433 83279 50288 41971 69399 37510 58209 74944 59230 78164 06286 20899 86280 34825 34211 70679 . . . and so on forever

Fig 10.3 *The beginning of the infinite decimal expansion of the number π. If the sequence is exhaustively random then all possible sequences of numbers will eventually arise in this infinite list.*

If this process could be implemented, then even more astonishing things could be achieved. Alan Turing, pioneer of computing, showed that there exist mathematical operations which cannot be carried out by any computer in a finite number of computational steps. They are called uncomputable operations and their existence is closely associated with the famous incompleteness theorem of Kurt Gödel, which teaches us that there exist statements of arithmetic that we can never prove to be true or false by using the rules of arithmetic. Uncomputable tasks cannot just be calculated by performing the same trick over and over again. They require something novel to be introduced at each step. Many uncomputable operations are known, and it is possible to prove that they are uncomputable Their defining characteristic is that a computer set the task of carrying them out would never stop.

But suppose that you had a computer that could carry out a super-task. Revolutionary new possibilities open up. Uncomputable problems could be solved in finite time. Better still, many other great unsolved problems of mathematics could be decided by explicit search of an infinite number of possibilities.

Take, for example, Goldbach's conjecture, first made in 1742. It claims that every even number can be made up from the sum of two prime numbers: for example 2 = 1 + 1, 4 = 2 + 2, 6 = 3 + 3, 8 = 5 + 3, 10 = 7 + 3, and so on forever. A few years ago the English publishing house Faber published a novel[16] whose hero spent his life searching for a proof of Goldbach's Conjecture. To boost interest in the book, the publishers offered a £1,000,000 cash prize for the first proof of the conjecture, or counter-example to show it is false. Alas, none has been found so far and the money is still safely in the Faber bank account.[17] Sometimes one hears the speculation that great unsolved problems of this sort might be examples of statements that are undecidable, as Gödel's theorem shows must exist. Remarkably, some statements can be proved to be undecidable, but Goldbach's is not one of them. Curiously, if it were undecidable we would have to conclude it is true. For, if it were undecidable then a systematic computer search of the sums of numbers that make up each even number could be made. If it were undecidable, then no counter-example could be found by such a search and so the conjecture would have to be true!

Let us imagine that computers capable of performing super-tasks would be able to decide the truth or falsity of conjectures involving uncomputable operations. If they could search systematically through all possibilities in a finite amount of our time, then they could print out 'true' or 'false' and stop. This is not as exciting to mathematicians as it might sound. Mathematicians are not only interested in whether conjectures like Goldbach's are true or false, they are interested in the forms of reasoning needed to prove it. They want to see new types of argument. A classic example was the proof of Fermat's Last Theorem by Andrew Wiles and Richard Taylor.[18] The truth of Fermat's conjecture emerged as a particular case of a much more general result that opened up types of proof and alternative formulations of old questions. A 'proof' by direct search would provide no new insights of that sort. It would, in effect, be like looking up the answer in the back of the book. In fact, if a conjecture like Goldbach's were shown to be true by an infinity machine, then we would feel aggrieved at being denied the insight provided by a proof. But if it proves to be false, then we are not really denied anything. The computer would never need to complete a super-task. The counter-example that is needed to show the conjecture to be false would be found after a finite period of search. It would be displayed before our eyes, just as it would by a human mathematician. Only if the human mathematician had constructed it by developing a deep insight into the nature of numbers would we be the loser by obtaining our counter-example from the computer search.

SOME NORSE CODE

'So put me on a highway and show me a sign
and take it to the limit one more time'

The Eagles, *Take it to the Limit*

Infinite sequences of events like the Thomson Lamp's on-off sequence can have bewildering properties. Suppose that we consider an unending sequence of switchings scoring +1 for 'on' and –1 for 'off'. Then, by adding up the alternating sequence of +1 and –1 scores, we can determine whether the lamp is off or on after any number of switchings.

If we start with the lamp on, then the sequence of on-off-on-off-on- . . . switchings gives us a sum that looks like

$$1-1+1-1+1-1+1-1+1-1+1-1+1-1+1-1+\ldots$$

If we stop the sequence after any number of switchings, then we can work out the total score. After any *even* number of switchings the total will be zero and the light will be off; but after any *odd* number of switchings the total will be +1 and the light will be on.

So, if we want to know whether the light is on or off after an infinite number of switchings all we have to do is find the infinite sum of the series. This is exactly the perplexing series that we encountered on p.65 in Chapter Four. Remember that we can group the terms in the series so that the sum is either 0, or 1 or even ½. In the case where the sum works out as zero, the light will be *off* after an infinite number of actions. When it comes to 1, then the light will be *on*. But the oddest conclusion arises when we break the series up as

$$S= 1-(1-1+1-1+1-1+1-1+1-1+1-1+1-1+1-1+\ldots)$$

Nothing odd about that, but remember what we learned in Chapter 4: the unending series in the brackets is exactly the same as our original series S; so we have

$$S = 1-S$$

And so S = ½ this time! This time the lamp is neither off nor on. It's half on, in a state that is like an average of the two.

These answers teach us something very important about infinite series of terms, as well as about infinite processes. The alternating series S is not a convergent series. Such series were once referred to by the great Norwegian mathematician Niels Abel (Figure 10.4) as

> 'an invention of the devil and it is a shame to base on them any demonstration whatsoever. By using them, one may

> draw any conclusion he pleases and that is why these series have produced so many fallacies and so many paradoxes.'[19]

It does not have a unique sum. We can only define its sum if we also specify the process used to enumerate it. This is never the case for the sum of a finite number of terms. Thus there is a way of working out the limit such that the light is 'on' and another such that it is 'off' after an infinite number of switchings. Most sobering of all though is the other lesson we learn from this example. If we had stopped the series after any finite number of terms, no matter how large, then the sum of them all would always have been 1 or 0. But our third method of summing the series gave us an infinite sum equal to $\frac{1}{2}$ which no finite sum could ever give.

There is something about the infinite sum that no finite part of it, however long, can give. The question that asks if the lamp is 'off' or 'on' after a minute is not meaningful. It has no answer.

Fig 10.4
Niels Henrik Abel (1802–29).[20]

THE END-GAME PROBLEM

'As it was in the beginning, is now, and ever shall be. World without end.'

The Magnificat[21]

There is a further perplexing problem about these infinity machines that seems to transcend the issues of practicality, of whether buttons can be pressed quickly enough, or whether consecutive actions of the machine can be physically distinguished from one another. It is the end-game problem. Suppose that an infinite number of tasks can be performed by our wonderful new limitless notebook computer in the space of the next hour. What could that mean? How would the super-task end? What is the last action of the computer going to be? Here is one philosopher's reaction to this dilemma:

> 'The difficulty, as I see it, is not insufficiency of time, tape, ink, speed, strength or material power, and the like, but rather the inconceivability of how the machine could actually finish its supertask. The machine would supposedly print the digits on tape, one after another, while the tape flows through the machine, say from right to left. Hence, at each stage in the calculation, the sequence of digits will extend to the left with the last digit printed being "at center". Now when the machine completes its task and shuts itself off, we should be able to look at the tape to see what digit was printed last. But if the machine finishes printing all the digits which constitute the decimal expansion of pi, no digit can be the last digit printed. And how are we to understand this situation?'[22]

There seems to be a logical impasse here. If an infinite number of tasks are completed in one hour, then it should be possible to inspect the last task performed at the end of the hour. But surely this could never be? Suppose our infinite task was a straightforward one, like printing out all the positive numbers: 1, 2, 3, 4, 5, 6, . . . and so on. There is no last number, so there could never be a final print-out by the machine.

In response, a super-task enthusiast could say that he is not claiming that every infinite listing can be performed by his infinity machine, and the listing of the positive integers that we have chosen as a test-case is one of those that can't be done. Just because one task fails does not mean that all others would fail.

At first, this sounds a reasonable counter, but it is not as telling as it might first sound. As we have seen, Cantor taught us that all basic infinities can always be 'counted' by putting them in a one-to-one correspondence with the positive numbers – that is why they are called 'countable' infinities. Not being able to write down the last positive integer means that you will not be able to print out the last action in any other countably infinite process. Infinities that are bigger than this countable number – for example all the numbers with never-ending decimal expansions – will fail too in this respect because they contain the countable infinities,[23] and much more besides.

The 'end-game' problem is a deep-seated conceptual problem that challenges the coherence of the whole concept of doing an infinite number of things in a finite time. Yet we have seen that it is by no means nonsensical to divide a finite interval into an infinite number of ever-shortening pieces. This is what the examples of Zeno and of Weyl challenge us to believe. So what is it that prevents an infinity machine from dancing to the same beat?

RELATIVITY AND THE AMAZING SHRINKING MAN

'be not ignorant of this one thing, that one day is with the Lord as a thousand years, and a thousand years as one day.'

St Peter[24]

Under closer inspection, Thomson's Lamp begins to look about as likely as Aladdin's. Infinity machines that *do* things face a raft of problems. One of the reasons that these machines were of interest at first as puzzles, like Maxwell's Demon, was that the old 'classical' physics of Newton appears to impose few limits on the functioning infinity machines like those proposed by Weyl. For, in Newtonian physics, there is no limit to the speed at which signals can travel: there is no limit to how fast switches can respond, signals can move.

A little dose of reality is needed at this point. Einstein taught us that there is a fundamental limit to the speed at which information can be transferred in Nature. There is a cosmic speed limit: the speed at which light moves in a perfect vacuum. This simple idea has many unexpected consequences and it underpins all that we know about the physical world.

In the world according to Isaac Newton, we can observe light to travel at many different speeds, just like anything else. Stand by the road side and shine a torch down the street. The light will move at a particular speed relative to you on the street. But what happens if a car drives past with its headlights on (see Figure 10.5)? If you were Newton, you would think that relative to you the light from the car would move at the speed at which it radiates from the light bulb (the same as the speed with which it shines from the torch) plus the speed at which the car is moving. Every passing car has a slightly different speed and so, relative to you, light from different cars will be moving at different speeds. In Newton's world-view there was no maximum speed for the movement of the light and no cosmic speed limit. The speed of light depends on its source.

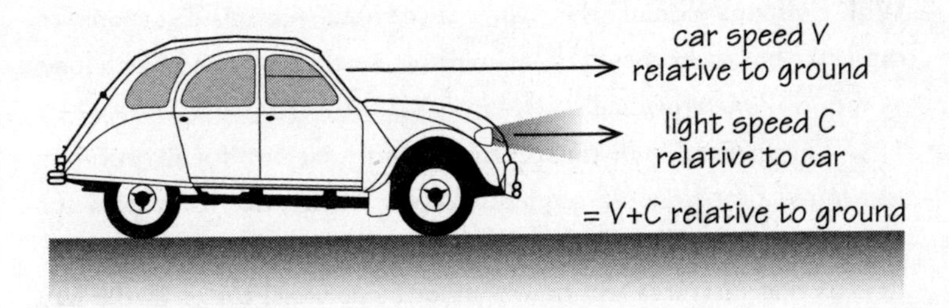

Fig 10.5 *In Newton's world view the speed of light shining from the headlights of a moving car has different values relative to the ground and relative to the car.*

Newton's theory was not conceived to deal with objects that move at or near the speed of light. Einstein showed that in order for the logic of cause and effect to remain consistent, everyone must measure the same speed for the motion of light, no matter what they measure it relative to. This is surprising. According to Einstein, if an observer moving at speed U in one direction relative to the ground launches a rocket with a speed V in the same direction relative to him, then the speed of the rocket in the same direction relative to the ground is not U + V as Newton would have believed, but

$$(U + V)/(1 + UV/c^2)$$

where c is the velocity of light. This is a remarkable rule. First, notice what happens when the speeds involved are far less than the speed of light: $U << c$ and $V << c$, so $UV << c^2$ and $1 + UV/c^2$ is approximately equal to 1. Einstein's formula for the relative speed becomes equal to U + V to a very good approximation: Newton's theory is the limiting situation of Einstein's theory when speeds of the objects considered are much less than that of light.

Now look what happens when we pick U and V both equal to c and ask what speed light appears to travel at when emitted by a source that moves at light speed. In Newton's theory the answer would be 2c. With Einstein's formula the answer is $(c+c)/2 = c$! And this answer can never be bigger than c no matter how we choose U and V, so long as they do not individually exceed c.

The universality of the speed of light in a vacum[25] for all observers is a foundation stone of modern physics.[26] The price to be paid for this universality of a speed, which is a rate of change of distance with time, is that distances and times cannot be universal things in the way that Newton believed. There can be no universal absolute time that everyone experiences and measures irrespective of how they are moving, no unambiguous measure of the length of anything that holds irrespective of the motion of the measurer.

If we measure the length of a rod to be L when it is not in motion relative to us, then we will not measure it to have the same length if it is in motion relative to us. If it moves past us at a constant speed V, then we will measure its length to be $L' = L(1-V^2/c^2)^{\frac{1}{2}}$.

Since V is always greater than zero but no greater than c, we see that L' will always be less than or equal to L. We will see the rod to be shorter than when we observe it at rest relative to us. This length that we measure when the rod is at rest with respect to us is the greatest length the rod could be found to have and it is called its 'rest length'.

This reduction in the observed length created by relative motion is called 'length contraction'. It means that there is no absolute concept of the rod's length that exists independently of our observation of it – hence, the caricature of Einstein's relativity theory as showing that 'everything is relative'. But remember that the speed of light is not relative. All observers measure it to be the same, irrespective of their motion or that of the light's source.

As with length, so it is with time. Suppose that we measure an interval of time recorded by a clock to be T when we are not moving relative to the clock. When we move with a constant speed V relative to the clock, we will measure the same interval of time to be $T' = T/(1-V^2/c^2)^{1/2}$. We see that T' is always longer than T: moving clocks are measured to go slow. The length of my life depends on the motion relative to me of the people who are measuring it. In everyday life the speeds of cars and other objects we encounter are so much smaller than the speed of light that these changes are imperceptible. However, watch the motions of fast-moving cosmic rays or particles moving in accelerators and these changes to space and time are routinely observed. The only unambiguous time is that recorded by a clock that shares your motion. What it records is called 'proper time'.

It is important to appreciate that these remarkable features of space and time are not just optical illusions created by fast motion in the way that an object might appear distorted in shape because we are viewing it at an angle, or because high-speed motion damages the clocks or distorts the rulers. The changes are real. For we must now recognise that time and distance are not the fixed things. We grew to believe that because our experience of motion was biased by studying what happens at speeds far less than that of light. Only when we could produce or detect motions close to the speed of light did these subtle changes become evident.

A MATTER OF TIMING

'Between the idea
And the reality
Between the motion
And the act
Falls the shadow

T.S. Eliot, *The Hollow Men*

Earlier, we highlighted the importance of actions generating entropy in order to count in tallying the total number of tasks completed in a finite interval of time. Steps in innocuous infinities generate no entropy. We can now consider a specific example of how including such spurious 'tasks' would bring about a violation of Einstein's principle that no signal can spread at a speed exceeding that of light in a vacuum.

We are going to consider an imaginary numbering process, familiar to anyone who has anything to do with military drills. Take a regiment of soldiers and stand them in a line, one behind the other. The first soldier shouts 'one' and on hearing him the soldier in front of him shouts 'two' and so on down the line. A number 'signal' will run down the line at a speed determined by their reaction times to the shout in their ear from behind. The number certainly couldn't go faster than the speed of sound and in practice it would go quite a lot slower. The speed with which the number signal passed from the start to the end of the line could never exceed the speed of light.

Now suppose that we looked at another way for the signal to get from one soldier to another. We equip each soldier with a radio receiver. If each soldier was to receive an outside signal whose arrival time was precisely pre-programmed, it would be possible for the individual responses of the soldiers to appear to travel down the line faster than the speed of light. How is this possible in the light of Einstein's dictum that nothing can travel faster than light?

What Einstein taught us was that no information – no signal – can travel faster than light. In the case where the soldiers each respond to the shout of their neighbour there is a signal running down the line. Each shout is triggered by the previous one. Information is transferred. But when the soldiers are each cued by an outside signal there is no information transfer from one soldier to another down the line[27] although it may appear that way to an outsider who is not party to what is going on. Actually, the outsider has simply observed a collection of completely independent events that have no effect on one another. There is no signal and no violation of Einstein's speed limit. The situation at any moment is not a logical consequence of its state at any earlier moment. No coherent super-task is being completed.

NEWTONIAN SUPER-TASKS

Mr Rogge memorably compared Greek [Olympic] preparations to the whirling Syrtaki dance in the hit film *Zorba the Greek*: 'It starts very slowly, it accelerates and by the end you can't keep up with the pace.'

Guy Alexander[28]

Our introduction of the strictures of Einstein's theory of relativity into the discussion of super-tasks is very important. For without them there are unexpected ways to build an infinity machine. Philosophers of science have worked hard over more than fifty years but never made contact with the most interesting area of physics that provides simple examples of infinity activity in finite time. To find them we need look no further than Newton's famous theory of gravity.

Newton's theory of gravity describes how collections of masses behave under the influence of a law of force that varies inversely as the square of the distance between each pair of particles. This sounds quite simple. But appearances can be deceptive. We have only ever been able to find an exact solution of Newton's equations when there are two masses in the problem. As the number increases above two the problem becomes extremely complicated and, except for very special configurations of particles, we are reduced to following their behaviour using powerful computers. The complexity derives from the fact that if we put three equal masses together in orbit then they will eventually kick out one of the particles and settle down to a more closely bound orbit of two particles. This gravitational 'slingshot' effect is actually rather useful. Spaceflight planners use it to boost the speed of spacecraft on planetary missions. By trailing a planet or moon in just the right way they can receive a gravitational 'kick' that will dramatically boost the rocket's speed and reduce the need to carry so much fuel. At a more down-to-earth level you can get a feel for the instability of a three-body problem by dropping a large ball – like a basket ball – to the ground from chest height at the same time as a small table-tennis ball which just touches its upper surface. The big ball hits the ground first and as it rebounds it strikes the small ball on the way down. The result is dramatic. The table-tennis ball will rebound nine times higher than it would have done if it had just rebounded from the ground!

As mathematicians have studied Newton's theory of gravity they have found that it has very strange properties. If we get together more than four masses then there are solutions of Newton's equations for which the largest separation between any two of the masses will increase faster than *any* rate you care to specify. In a world governed by relativity the separation cannot grow faster than in direct proportion to the time. This means that systems of masses can become infinitely large but it was believed that they would always need an infinite time to do so. In 1971, Jeff Xia of Northwestern University made a dramatic discovery.[29] He showed that systems of more than four masses governed by Newton's law of gravity were permitted to become infinitely separated in a *finite* time.

Xia's archetypal example is shown in Figure 10.6. We have four equal masses that form two double systems orbiting with equal but opposite rotation speeds, so that the overall rotation is zero. Their orbital planes are parallel to each other. Next, Xia introduces a lighter particle that oscillates back and forth along the line of centres drawn between the two orbiting pairs. Each time the small mass encounters the influence of one of the two pairs of heavier particles it creates a little three-body situation and gets kicked strongly back, just like our table-tennis ball, while each of the orbiting pairs orbit slightly closer. Xia showed that this process continues back and forth and the pairs of particles get farther away from each other while the small mass oscillates back and forth between them at ever-increasing speed. Remarkably, the maximum distance between the particles becomes infinitely large in a *finite* time. The only consolation is that the starting conditions needed to achieve this result are extraordinarily unlikely. This dramatic behaviour is possible whenever there are more than four particles feeling the force of gravity but it is not known whether it is possible with only four.

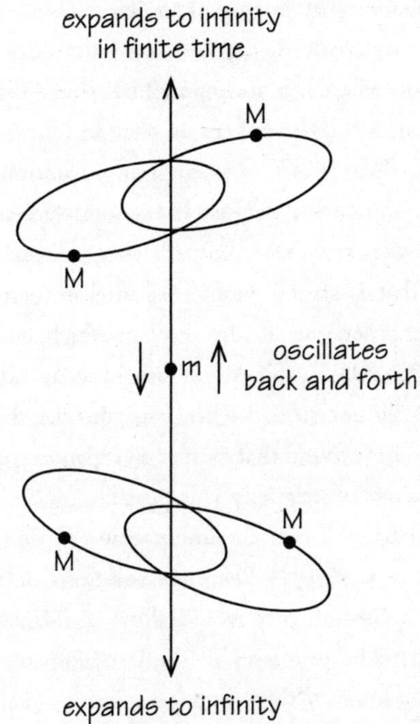

Fig 10.6 *Xia's configuration that goes to infinity in a finite time, executing an infinite number of physically distinct oscillations in the process. Four particles of equal mass, M, reside in two binary pairs rotating in opposite directions with equal speed. A fifth, lighter, particle m oscillates between the two binary pairs at ever increasing speed as it gets evicted by each of the pairs in turn. Each time it gets evicted it moves with increased speed and the pairs each become a little more tightly bound by the force of gravity. The system becomes infinitely large in a finite time and the lighter particle performs an infinite number of oscillations in a finite time.*

This example enables us to conceive of an infinity machine. The 'machine' is simply provided by the oscillations of the light fifth particles between the orbiting pairs. It will oscillate back and forth an infinite number of times in the finite period of time before the separations of the orbiting pairs becomes infinite. An infinity machine *is* possible in a Newtonian world, albeit with starting conditions that have virtually a zero probability of arising naturally.

Einstein's relativistic theory of gravity does not allow behaviours like this to occur. There is a maximum 'kick' that a three-body problem can exert on one of its members: it can't be evicted faster than the speed of light. There is also a maximum gravitational force that one body can exert on another because they cannot get arbitrarily close to one another. If they try, then eventually they will produce a local gravitational field that is strong enough to envelop them both in a black hole. This is another one of the ways in which black holes act as a cosmic censor (pp.105–7). Black holes might seem bad but, like growing old, they are really not so bad when you consider the alternatives.

Recently, I have found that even in an expanding universe described by Einstein's theory of relativity it is possible for a singularity of infinite pressure to arise all over the universe at a finite time in the future while it is still expanding.[30] There are solutions of the theory which permit this to occur and they would allow an infinite number of bits of information to be processes in the last moments as you run into the singularity. However, these strange solutions require matter to be able to transmit information at unlimited speed. If we impose the cosmic light-speed limit then this sudden end to the universe cannot happen.

RELATIVISTIC SUPER-TASKS

'Tragically, I was an only twin.'

Peter Cook[31]

These features of the world create new interest in the whole question of super-tasks. Could it be that one moving observer could see an infinite number of computations occur, even though only a finite number had occurred according to a programmer not moving with respect to the computer (Figure 10.7)?

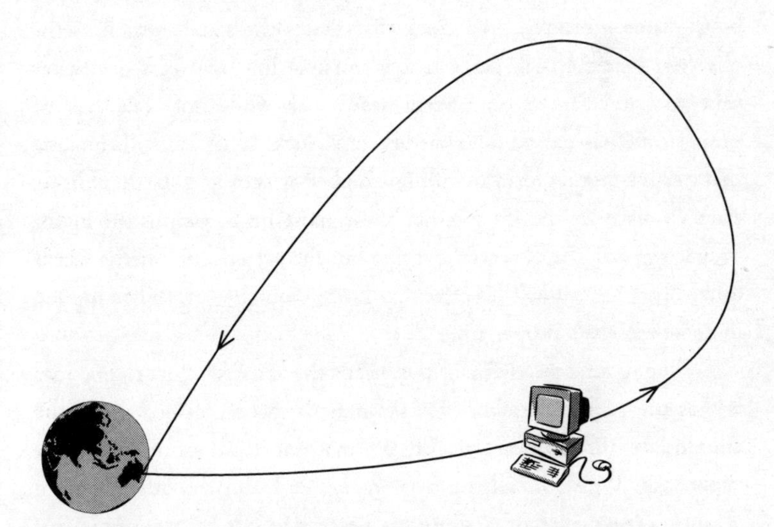

Fig 10.7 *Can we send a computer on a trip that will permit it to accomplish an infinite number of calculations in a finite period of time?*

The famous motivating example of relativity theory is the so called 'twin paradox'. Two identical twins are given different futures. Tweedledee stays at home while Tweedledum goes away on a space flight at a speed approaching that of light. Tweedledum eventually decelerates, turns around and returns home to be reunited with Tweedledee. Relativity predicts that Tweedledum will return to find Tweedledee much older than himself. The two twins have experienced different adventures in space and time because of the acceleration and deceleration that Tweedledum had to undergo on his round trip. We can imagine some more extreme version of this process where the stay-at-home twin appears infinitely old to the returning travelling twin.

The idea of a super-task is no longer quite so clear cut. If we focus on the amount of proper time that passes – this is defined to be the time measured by a clock that shares the same motion as the observer – in the twin paradox it is possible for 100 years of proper time to have elapsed on Tweedledee's clock while only one year of proper time has passed on Tweedledum's clock. If we are still thinking that a super-task requires an infinite number of actions to be completed, then we have to specify whether these must be accomplished in the proper time of the observer carrying out the actions, or whether some other observer can just have seen an infinite number of things happen in someone else's proper time.

The natural requirement that keeps the spirit of the original idea is that the infinite number of tasks must be accomplished in a finite amount of the proper time of the machine that accomplishes the super-task. We shall call this a *proper super-task*. On the other hand, if an infinite sequence of actions can be carried out by a machine in a finite interval of another observer's proper time, then we shall call that achievement a *pseudo super-task*. All proper super-tasks are also pseudo super-tasks, but not all pseudo super-tasks are proper super-tasks[32] (Figure 10.8).

In the old world picture of Newton that still holds good to high

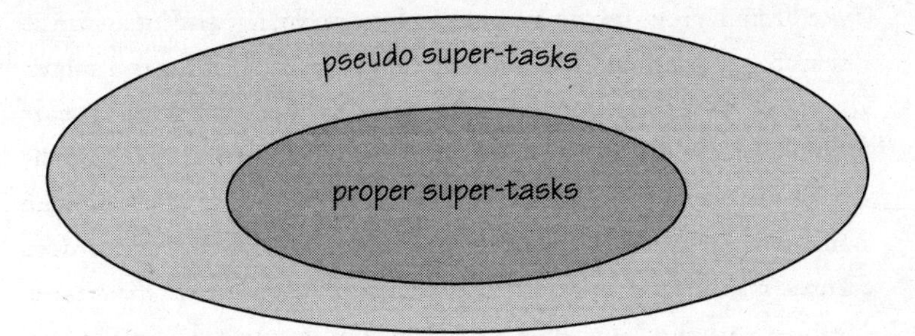

Fig 10.8 *Proper super-tasks are a subset of all pseudo super-tasks.*

accuracy when motions occur at speeds far smaller than that of light, there is no difference between proper super-tasks and pseudo super-tasks.

It was first pointed out by Itamar Pitowsky of the Hebrew University of Jerusalem that if Tweedledum can accelerate his spaceship sufficiently strongly, then he can record a finite amount of the Universe's history on his own proper time clock whilst his twin brother, who is not accelerating, records an infinite amount of proper time elapsing on his clock. A pseudo super-task seems to be possible in principle without doing violence to the structure of space and time and the laws of relativity.[33] Pitowsky wanted to know if this device would permit a 'Platonist computer' to exist – one that could carry out an infinite number of operations along some trajectory through space and time and print out an answer that we could see. Alas, in this simple example the observer who measures the infinite history cannot have access to the information that it contains.[34] It cannot reach him. In order for the receiver to stay in contact, he has to accelerate dramatically as well in order to maintain contact with the flow of information. Eventually the g-forces become stupendous and he is torn apart, no matter what he is made of.

This is a common obstacle that arises in many simple attempts to exploit the existence of pseudo super-tasks to solve infinite problems in a way that would render them effectively proper super-tasks. Here is another particularly graphic example. Suppose that Tweedledee and

Tweedledum grow up to become ambitious young mathematicians, desperate to determine the truth or otherwise of Goldbach's famous conjecture (p. 222). Tweedledum becomes fanatical in this quest and decides to sacrifice himself so that they can learn the truth. He takes a trip in a spaceship which he steers towards a black hole. The gravitational pull of the black hole will draw him inexorably in and steadily increase his acceleration as he falls towards the singularity at the centre. He knows that it will only take a finite amount of his proper time before he hits the singularity and is crushed out of existence. Meanwhile, Tweedledee is watching. He should see an infinite amount of his own proper time elapse before Tweedledum is destroyed. This is not only comforting in a fraternal way, but it also allows him to see an infinite number of his brother's computer calculations. This will tell him whether Goldbach was right or not.

Alas, black holes have a defence mechanism. The horizon of the black hole prevents Tweedledum's information reaching Tweedledee back on the outside. The 'cosmic censor' doesn't like super-tasks.

Again in this problem we see that if the twins had both fallen through the black hole horizon, then, although one might have been able to send the required information to the other, they would both end up being torn to pieces by the tidal forces of gravity as they fell close to the centre of the black hole.

Despite these problems in these simple cases of black holes and accelerated travellers, it has been discovered that the curved space-time geometry of Einstein's general theory of relativity *does* permit the existence of proper super-tasks which are not pseudo super-tasks. They have become known as Malament-Hogarth (MH) universes after the University of Chicago philosopher David Malament, and a Cambridge research student Mark Hogarth,[35] who discovered their possible theoretical existence in 1992.

There are solutions of Einstein's theory of general relativity which describe what MH universes do. Unfortunately, they seem to have properties that suggest they are not physically realistic. In particular

- The future is not uniquely and completely determined by the state of the universe at the present time in all MH universes.
- Time travel occurs in some MH universes. While this does not render them impossible, it is problematic.
- Some inhabitants of these universes will find that any amount of radiation, no matter how small, is compressed to zero wavelength and infinite energy. Any attempt to send the output from an infinite number of computations will zap the receiver and destroy them.
- The 'computer' that is required to carry out the infinite number of computations must be infinitely large. If we want to store an infinite amount of output we need somewhere to put it.

These are dire problems and they seem to rule out the practicality of engineering a relativistic machine in the laboratory that might carry out an infinite number of tasks in a finite time in a way that would enable us to receive and store the information without being destroyed in the process. Only one type of scenario has been imagined where the computations could be accomplished.

BIG BANGS AND BIG CRUNCHES

'If it were done when 'tis done, then 'twere well
It were done quickly.'

William Shakespeare[36]

The big problem with Zeno's Paradoxes is that the infinite subdivision of a finite period of time that he proposes does not correspond to a sequence of physically distinct operations. There is one situation where the paradox of Zeno might be made into a reality. Cosmologists have always been interested in trying to understand the complexities of the beginning of the expansion of the Universe and, if the Universe eventually reverses into contraction, what happens as it approaches the Big Crunch (see Figure 10.9).

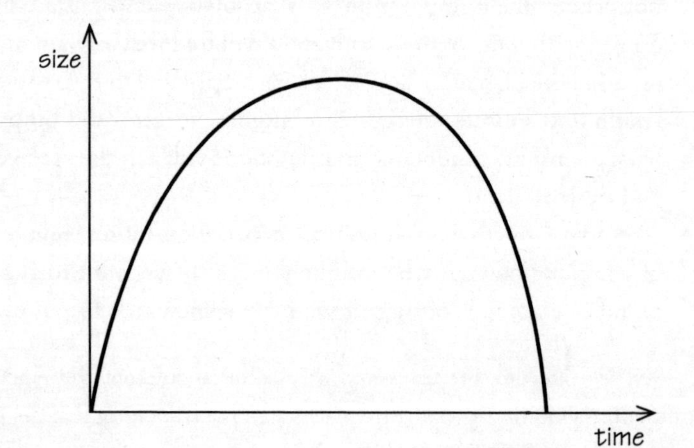

Fig 10.9 *A universe that expands away from a Big Bang before contracting back to a Big Crunch.*

The simplest imaginable universe like this begins expanding at the same rate in all directions at some finite time in our past, which we will label time-zero, and contracts back to a crunch at some finite crunch-time in the future. In reality we don't expect universes to expand at exactly the same rate in every direction, and when they become asymmetrical like this they behave in a very complicated way. Although they expand in volume, one direction tends to contract while the other two expand, tending to create an expanding 'pancake'. But soon the contracting direction switches to expansion and one of the other two expanding directions switches into contraction. Over a long period of time, the effect is a sequence of oscillations which proceed in a random permutation of the expansion rates in the different directions. This behaviour was discovered in 1969 by an American physicist, Charles Misner, who dubbed it the 'Mixmaster' Universe, after a well-known American food-mixer![37]

The striking thing about the sequence of oscillations of the volume of the universe as it shrinks to zero, when one runs its history back into the Big Bang at time-zero, or on into the Big Crunch at crunch-time, is that an infinite number of oscillations occur. This is rather like trying to draw the graph of x^2 times the sine of $1/x$ as x approaches zero (Figure 10.10).

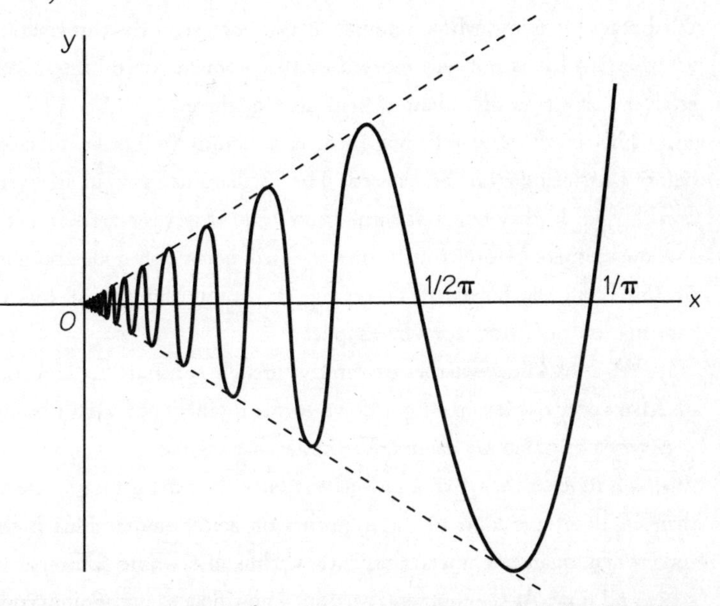

Fig 10.10 *The graph of x^2 sin $(1/x)$ versus x as x approaches the value zero. The graph must possess an infinite number of oscillations on any finite interval of x around x = 0. This is impossible to draw and so we have shown just a few of them.*

There would be an infinite number of oscillations of the graph – impossible to draw, of course.

The difference between the Mixmaster Universe and Zeno's Paradox is that an infinite number of physically distinct, real events happen in any finite interval of time that includes time-zero or crunch-time.[38] Measured by a clock that 'ticks' on this oscillatory time, the Mixmaster Universe would be judged to be infinitely old, because an infinite number of things have happened to the past in this time, and it will 'live' forever because an infinite number of things are still to happen in the future.[39] If a universal computer existed which processed 'bits' of information every time there was an oscillation in the expansion of the universe between two directions, then this universe would do an infinite amount of processing. It would complete a super-task. But because it is the whole universe, it does not send the information anywhere and it is not encumbered by the problems of infinite accelerations, or bursts of radiation. It is all that there is.

This is all very well, but there is a significant hurdle to cross before this infinity can be realised. The oscillations grow in size very quickly and, if they begin at some finite time after time-zero and end at some moment before crunch-time, they will occur only a *finite* number of times. All the hopes of processing an infinite number of bits of information or 'living' forever disappear.

We don't know whether we can legitimately continue the sequence of Mixmaster oscillations right down to the moments of zero volume. It is very likely that we cannot. We know that the nature of space and time will have to change in a radical way once the expansion gets closer than 10^{-43} of a second to the moments of zero volume. This is the time when quantum uncertainty overwhelms the whole Universe of space and time. At the moment, we don't know how to carry on making predictions about what will happen closer to the Bang and the Crunch than this. And, even though this breakdown time is fantastically small, in a Universe like ours it is enough to reduce the number of Mixmaster oscillations from infinity to about a dozen.

The original Mixmaster Universe had a finite future. Universes that expand forever seem to have a very different fate, getting cooler and emptier, devoid of the habitats for life to evolve and survive. There has grown up a widespread belief that life may inevitably have to die out in universes that expand forever. It is hard to second guess what life might be like or do, but we could try to identify the most basic necessary attribute of 'life' – the ability to process information – whatever its form.

The speculative cosmological futures' market has two visions to offer. If our Universe will continue to expand forever, slowed only by the decelerating pull of gravity, then there is hope for information to continue to be processed forever. Sigbjorn Hervik and I have shown that the differences in the expansion rate of the Universe from one direction to another can be used to create differences in the temperature of radiation moving in different directions (Figure 10.11).[40] These differences can then be used to drive a computational process. An infinite

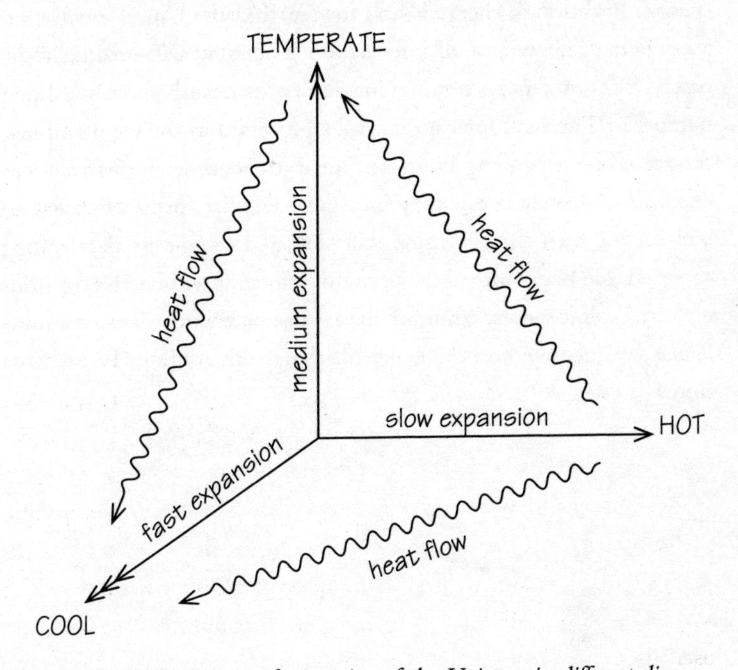

Fig 10.11 *Different rates of expansion of the Universe in different directions are inevitable. They can be exploited to make radiation cool at different rates in different directions. The resulting temperature difference can then be used to drive 'machines'. Energy is thus extracted from the expansion-rate variations in the Universe. After it has been extracted the expansion will be a little more symmetrical than before in order to conserve energy. Curiously, 'life' is driven by tidal power, although it is the tidal power provided by the gravitational-wave tides in curved space.*

number of distinct operations of this process can occur in the infinite future lifetime of the Universe.

The alternative future arises if the Universe's expansion continues to accelerate forever. At present, we see that the expansion is accelerating. This is somewhat mysterious. We know that this happens if an unusual type of matter exists in the Universe which exerts gravitational repulsion on itself and all other forms of matter. We know nothing more about its specific identity and have taken to calling it the 'dark energy'. If this dark energy is here to stay, then the Universe will keep accelerating forever and all information processing will eventually die out.[41] 'Life' of any sort will eventually be extinguished: only a finite number of bits of information can be processed in the whole infinite future of the expanding Universe. But if the dark energy is transient and will ultimately decay away into more familiar forms of radiation and energy, then the expansion will slow and resume its decelerating career. If life has managed to survive that long, it will be able, in principle, to exploit the slackening of the expansion rate in order to continue 'living', calculating, and changing. All things can continue to be made new.

chapter eleven

Living Forever

'Millions long for immortality who don't know what to do with themselves on a rainy Sunday afternoon.'

Susan Ertz[1]

CHILDHOOD'S END

'And even if we were to succeed in imagining personal immortality, might we not feel it to be something no less terrible than its negation?'

Miguel de Unamuno[2]

Living forever promises to be a long-winded business. Yet it has been the dream of mystics and the promise of religions for thousands of years. Human life-expectancy has not changed very significantly since records began. We might now live two or three times longer than the oldest Neolithic humans, but that is as nothing on an infinite scale of things. The longest human life we know of lasted 122 years, but in recent years, specialists in gerontology are talking about genetic control of ageing that might conceivably double life-expectancy within the lifetime of some people alive now.

As soon as we start to think like this, there is no end to the number of years that we might, in some sense, 'live' for. If all our organs can be replaced by spare parts from others, or from spare-part factories, then our bodily evolution becomes more like that of a cherished classic car whose parts are continually replaced until we realise that there is nothing left of the original. This idea should no more have a negative impact than does the replacement of limbs or organs for disabled patients today. It emphasises that people are more than their external bodily appearance. Just as the computer program or image that you run on your computer transcends the machine and could be run on other machines, so it is with the self that is held within our minds. It owes much to our bodies, but it is far more than the collection of atoms of which they are composed.

The weight of opinion over the centuries seems to be that living forever would be a 'good' thing that we should strive for, but the reasons for such a belief were so diverse that it is hard to avoid the conclusion that they might cancel each other out. There were those whose lot on Earth was so bad, so ruined by injustice, poverty, and disease, that the hope of an infinite life to come made all the misery and unfairness of this brief finite life bearable. There were others who believed that they had already lived, and would die and be reborn, over and over again, forever. Others still, saw death as a welcome escape from the miseries of life or an incentive to make the most of life: to act carefully and responsibly with the time allotted for our lives. The prospect of a finite life was thought by many philosophers to be an incentive to use our time on Earth wisely. If life were unending, then there would be no natural development, no urgency, no sense of completion.

At the most mundane level, we can ask a biologist: what is the function of death in populations of living things? We read a lot about the search to understand the origin of life, but the origin of death is rarely discussed scientifically. Its role is clear, even if the exact sequence of events that brings it about is not completely understood. It creates diversity within a species. If no more offspring appeared, and no more deaths occurred after a population reached a certain level, than that population would be trapped by the sum total of its own genetic information and ingenuity. Innovation would slow dramatically and the population would ultimately be at a disadvantage compared with other less-long-lived innovators whose gene-pool was constantly being reshuffled and extended. Needless to say, if death by natural causes stopped but births continued, then the environment might start to get rather crowded and death by unnatural causes would ensue. The food chain might develop a lot of very weak links as well.

Death is also a good way to protect life from extinction from lethal epidemics. This might sound odd, but consider the way in which a lethal disease develops. If it kills its victims quickly, then the virus dies with them and has less chance of being spread. If a disease causes serious debilitation but not death, then it would be able to spread and dramatically weaken all the members of a population of members who lived forever.

Still, once medical science is sufficiently advanced, maybe we won't have to worry about disease and there could be other artificial means of sustaining genetic diversity. So let's leave these more biological issues in order to think about some of the oddities of a life like our own that has no end.

THE SOCIOLOGY OF ETERNITY

> 'I don't want to achieve immortality through my work. I want to achieve it through not dying. I don't want to live on in the hearts of my countrymen. I would rather live on in my apartment.'
>
> Woody Allen[3]

Many traditional religions place great emphasis on understanding and accepting death. The possibility of its removal leads to many unexpected issues. Imagine the Citizens' Advice Bureau being besieged by people considering whether to subscribe to everlasting life, or reporting the unforeseen problems that have arisen since they did so, after being taken in too readily by the sales person's late-night promotional phone call.

What sort of problems would the hapless adviser be facing in this fantasy world? Criminology develops a fascinating speculative future. Perhaps there would eventually be an upsurge in honesty and a dramatic drop in crime and dishonesty because the chances of it remaining undetected become so small when there is forever to find out the truth.[4] But if you are detected, sentencing becomes more complicated. What is meant by a 'life' prison sentence now? Can it really be forever in extreme cases? Where it is for a fixed term, what is the appropriate period? At present, we can think of twenty-five years as a fraction of an expected life span, but any finite number is a zero fraction of infinity. Patience becomes a real virtue for the criminal, as in other walks of life. If you plan for the far future, then the chance of an investigation linking events at two different times diminishes as the time between them increases.

As always there is much work for lawyers. If you suffer from someone's negligence, how is the accidental loss[5] or impairment of a human life to be valued for the purposes of compensation? At present the death or serious disablement of a person will lead to compensation being paid to family members. The amount is calculated so as to compensate for the loss of expected earnings. What happens when they are infinite?

Does anyone retire? If birth rates are reduced to accommodate the negligible death rate, then will such a society become increasingly more conservative, with everything run according to the experiences of the past, or will it become wildly experimental, seizing the chance to try out everything, safe in the knowledge that there is always time to put it right if it all goes pear-shaped?

What would be the future of marriage? Would polygamy become popular? Will family arguments lead to a gradual fracturing of all family ties over very long periods of time? Will extended families just dissolve because of the huge numbers of members? Is fraternity doomed? Families will surely have less and less significance as time goes on. Will this lead to more or less peaceful coexistence? At first sight, one thinks that relations will be better because of the family connections, but we know that a large fraction of violent incidents are intra-family.

What happens to those religious faiths that promise eternal life? Do they simply refocus on the quality rather than the quantity of life? What else will they offer adherents who do not fear death? For some religions, an eternal life in our present state would not be seen as a virtue. This life is a preparation for greater things to come. To be locked forever in the antechamber of paradise would be a condemnation, not a success. Perhaps one unexpected response to large numbers of people coming to regard living forever as a dreadful thing might be a new type of religion that promises only finite life in the future, a way of bringing life to an end in an ethically acceptable way. The 'second coming' would still be sought for, but its role would be strangely reversed: to end eternal life rather than to usher it in. Its role would have to be to transform the quality of life.

There would be a split in society between the super-achievers, those who respond to the expectation of living forever by trying to do everything, and the sub-achievers, those who do nothing because there is plenty of time to do everything, later on (whose attitude is described by a word like manãna, but which does not convey the same sense of urgency).

Alan Lightman, in his book *Einstein's Dreams*, identifies these two personality types and dubs them the 'Laters' and the 'Nows'.[6] The world they inhabit is strangely polarised. The Laters take eternal life slowly, for

> 'The Laters reason that there is no hurry to begin their classes at the university, to learn a second language, to read Voltaire or Newton, to seek promotion in their jobs, to fall in love, to raise a family. For all these things, there is an infinite span of time. In endless time, all things can be accomplished. Thus all things can wait. Indeed hasty

> actions breed mistakes. And who can argue with their logic? The Laters can be recognized in any shop or promenade. They walk an easy gait and wear loose-fitting clothes. They take pleasure in reading whatever magazines are open, or rearranging furniture in their homes, or slipping into conversation the way a leaf falls from a tree. The Laters sit in cafés sipping coffee and discussing the possibilities of life.'

In complete contrast, the Nows are driven individuals. Their distance from the Laters grows ever bigger as they achieve more, racing each other to see who can do the most, spurred on perhaps by intellectuals who tell them about Cantor's different orders of infinity, seeing in them a way to do infinitely more than other immortal super-achievers. Lightman reflects that

> 'The Nows note that with infinite lives, they can do all they can imagine. They will have an infinite number of careers, they will marry an infinite number of times, they will change their politics infinitely. Each person will be a lawyer, a bricklayer, a writer, an accountant, a painter, a physician, a farmer. The Nows are constantly reading new books, studying new trades, new languages. In order to taste the infinities of life, they begin early and they never go slowly. And who can question their logic?'[7]

Family life for the eternals is a cumulative dilution towards an all-encompassing equilibrium state. Infinite life means

> 'The Nows and the Laters have one thing in common. With infinite life comes an infinite list of relatives. Grandparents never die, nor do great-grandparents, great-aunts and great-uncles, great-great-aunts, and so on, back through

> the generations, all alive and offering advice. Sons never escape from the shadows of their fathers. Nor do daughters of their mothers. No one ever comes into his own. When a man starts a business, he feels compelled to talk it over with his parents and grandparents and great-grandparents, ad infinitum, to learn from their errors. For no new enterprise is new.'

The trend is clear. Marriage results in a never-ending sequence of in-laws to offer advice. Craftsmen never emerge from their apprenticeships. Engineers never finish great projects because there is no end to the considerations that must be taken into account, the wealth of experience that must be consulted. No one has the confidence to strike out on their own because someone has generally done it before. The world clogs up with unfinished projects, slowed by ceaseless referrals. Personal fulfilment is hard to find. Not only are there innumerable voices of experience whispering in your ear, but most of them are your relatives.

Suspicions are also rife. Secrets are hard to hide. They eventually all come out and marriages rarely last very long. It turns out, paradoxically, that fewer are undermined by the emergence of unsavoury secrets than by the very thought that such a future fate is inevitable. The same dissolution awaits all friendships. Those who desire close friendships become isolated as they fail to find any that last. Others amass thousands of ephemeral acquaintances, none of any lasting value. The psychological pressure imposed on everyone by the realisation that the possible has become the inevitable changes and diminishes the quality of life for them all. People begin to ask whether a finite life of vitality and fulfilment might not amount to more than an infinity of diminishing returns. Suicide becomes common for, in the words of Miguel de Unamuno, it is

> 'only he [who] desires personal immortality who carries his immortality within him. The man who does not long

passionately, and with a passion that triumphs over all the dictates of reason, for his own immortality, is the man who does not deserve it, and because he does not deserve it he does not long for it.'[8]

THE PROBLEM-PAGE OF THE UNENDING FUTURE

'A lawyer's dream of heaven: every man reclaimed his own property at the resurrection, and each tried to recover it from all his forefathers.'

Samuel Butler[9]

There is a play by Karel Čapek, that was then made into an opera by Janaček, which tells the story of a woman, Elina Makropulos, whose father was physician to an Emperor in sixteenth-century Europe.[10] Her father has created the Elixir of Life and decides that his daughter will be the subject of its first drug trial.

The Elixir works but, like any drug, it needs to be taken regularly if its effects are to persist. Elina has dutifully taken the medicine and has lived for 342 years. The play reveals that this huge lifetime has reduced her to a state of bored desperation, indifferent to the world around her, with nothing to live for. The friends of her youth have long since died. She refuses to take the Elixir again and dies. Others destroy the secret of the Elixir of Life, despite the protests of some of the more aged members of society.

This play inspired the English philosopher Bernard Williams to consider more carefully whether living forever was a blessing or a blight. Along with most philosophers who have considered this question dispassionately, ignoring traditional religious beliefs and expectations, he thinks that living forever would be a poisoned chalice. Although he would want to extend any finite period of life further, he sees the prospect of a never-ending life as a dismal future of repetition, boredom, and déjà vu. Reluctantly, he concludes that, despite his desire to continue living,

> 'an eternal life would be unlivable. In part as Elina Makropulos's case originally suggested, that is because categorical desire will go away from it . . . I would eventually have had altogether too much of myself. There are good reasons, surely, for dying before that happens. But equally, at times earlier than that moment, there is reason for not dying. But as things are, it is possible to be, in contrast to Elina Makropulos . . . lucky in having the chance to die.'[11]

Yet the implicit assumption that intellectual activity will exhaust all that is knowable in an infinite future is not necessarily a convincing one. We do not know whether the laws and structures that the Universe contains and can give rise to are finite or infinite in number.[12] And, if infinite, we do not know what order of infinity they might possess.[13]

We know, for instance, that mathematics is infinite in extent – there is no end to the number of new structures that might be generated from those that are known – but we don't know if they will continue to be interesting in the sense of containing novel features that are not just new examples of older ones. One suspects that novelty will not disappear as mathematics is mined out, because almost every mathematical statement is a Gödel undecidable one that cannot be resolved by any computer program, even if it runs forever. In this vein, Stephen Clark tells us of the imaginative immortals in Arthur C. Clarke's story *The City and the Stars,*[14] where

> 'the great city Diaspar, last and greatest of all human cities, is inhabited by perennials whose (edited) memories reach back a thousand million years. Some movement is allowed by the device that every such perennial steps back into computer storage after a thousand years of active life, to be reissued later. But any real youngster must have great difficulty finding any proper place in such a society, and far more difficulty . . . in changing it . . . Clarke's immortals are still working out the structure of prime numbers, creating works of art, exploring fantasies . . . Not all eternal cities are dull; even ones that do not much change may be worth having . . . The desire to live forever is the desire never to be ended or closed off: the desire, in effect, to contain everything, so that there is nothing outside oneself that one will not eventually grasp.'[15]

We might also ask what it is that enforces our finiteness upon us most strongly. Is it the inability to *do* everything possible? Or is it that we are aware of a sea of possibilities for us which are going to be shut off by death? Is it that there are places that we cannot reach, or simply that there will be people we know who we will cease to know? Is it that our curiosity will be quenched: that there will be things to know that we will not know? Or is it that death is just a 'bad thing' in itself, as the Christian tradition would have it, and thus something to be overcome?

Bernard Williams's meditation on the case of Elina Makropulos convinces him that death is a bad thing, because it closes off possibilities that would otherwise be open to us. None the less, immortality should not be preferred to mortality – at least if we retain our present human nature[16] – because mortality imbues life with its most important goals. Thus, although at any moment there is good reason to try to live longer, there is no reason to continue living forever. This dichotomy is similar to some of the features of infinite series that we have encountered in previous chapters. We have seen that it is possible for the sum of an infinite number of terms to have a property that is not shared by any member of the series. Williams's dichotomy is not dissimilar in its jump to a negative conclusion, despite all that has gone before.

All these evaluations of the pros and cons of living forever that we find in works of philosophy are similar in one interesting respect. There is never a mention of anyone but oneself. Unending life is entirely about self-gratification[17] and the need to provide a meaning for oneself through continuing to do things, think things, and look forward to the future. Elina Makropulos didn't look to a future of helping other people.

THE STRANGE, FAMILIAR, AND FORGOTTEN

'Memories,
Like the corners of my mind
Misty watercolor memories
Of the way we were'

Barbra Streisand[18]

Living forever with or without memory of the past – what is best? Much depends on whether those memories are good or bad. If memory is finite and fallible (as it is now), then perhaps living forever is just like being reborn. Eventually, you forget almost everything that happened to you before a certain date in the past. There is a past horizon, just as there is for us when we reach back to our earliest childhood memories.

Perhaps there are interesting variations in which there is competition for memory space between old memories and new ones. It is like having a computer disk, on to which you want to save information. If you are told the disk is full, you have to delete something to be able to save the new document. Individuals would have to decide what to forget in order to be able to remember new things tomorrow. Life would become like a chronic medical condition which requires daily treatment, with memory deletion applied while you sleep, or the next day would be literally a 'blank'. You would probably go for larger, less frequent, deletions of data, which you could always download on to CD for later use. Indeed, in the far future people may be far less possessive about memories. In the ancient world, before almost everyone could read, memory was vital. It was all you had to keep information about your family, your society, and their traditions. Even thirty years ago, there was still a considerable premium placed upon remembering information in the schoolroom. Today, the need to remember has never been smaller. The world wide web allows information to be retrieved in seconds. Perhaps in the far future we will have become even less self-sufficient in memory and recollection, realising, in the face of an avalanche of information, that we are fighting a losing battle. Or maybe that web will draw us all so close together, link us so irreversibly, lead us to be so similar in what we know, that we will have evolved into a new form of collective life. In Olaf Stapleton's classic novel *Star Maker*, the future leads to a single cosmic self in which all individuality, and all its traits like death and individuality, are effaced. One dies only if all die.

This making of back-ups of memory immediately suggests a yet more radical course. Why back-up only memories? Why not guard against death by making frequent clones of yourself? All you would lose would be the experiences that you had since the last back-up was made. The sensationalist press would find the whole thing endlessly titillating and there would always be the odd absent-minded professor who had forgotten to back himself up for thousands of years, and so was having to begin again with the knowledge of a teenager in the body of an ancient. And in a world where freedom of choice might be paramount, there would be some who would elect to begin again. Weary of their lot, they could choose to be reborn with any desired measure of their present attributes preserved in the new genetic materials. There would undoubtedly be whole scientific specialisations devoted to determining the consequences of preserving or over-writing.

These fantasies are surprisingly easy to concoct. One of the lessons one learns from trying to do so is how more readily one is drawn to consider the psychological aspects of living forever than the technical advances that one would exploit or the knowledge one would gain. There could be some very small print on the contractual agreement with the *Eternal Life Leisure Company*. Maybe time itself gradually slows down. So, even though you have an unlimited future time in which to do things, it takes longer and longer to complete them. It is like walking through thicker and thicker treacle. You might be told that you have only a finite number of heartbeats to experience in the future, but you can choose how slowly or how rapidly you take them.

Or maybe hibernation is the future. You need to sleep for longer as the days go by, because energy conservation is required. At present we sleep for approximately one-third of our lives. Suppose this fraction needed to increase, so that we had to hibernate for a larger and larger fraction of each year (sleep one year for each 100 years awake, then for each 10 years awake, then for each year awake, then each tenth of a year and so on). Be awake for a year, asleep for ½th of a year, awake for ⅓th of a year, asleep for ¼th of a year, awake for ⅕ . . . and so on forever.[19] We will live forever despite the ever-diminishing returns.

Alas, we have seen that our Universe seems to be set on a course which will see it accelerating away from the clutches of gravity in the future. Although we might like to contemplate the niceties of eternal existence, the prospects do not look promising in our Universe. The environment will become steadily more hostile. Energy will become harder to mine. The visible part of our Universe looks increasingly like an oasis where things are hospitable for just a while. Once there was no life in our Universe, and to this lonely lifeless existence it may one day return:

'Be comforted, small immortals. You are not the voice that all things utter, nor is there eternal silence in the places where you cannot come.'[20]

INCESTUOUS TIME TRAVEL

'if we could travel back to Cellini's Florence and *vice versa*, we would be appalled by the smells and he by the noise.'

Michael Dibdin[21]

Time travel introduces another form of living forever.[22] If time is a closed loop rather than a line, then history need have no end and no beginning. The possibility of travelling through time is much discussed, both in the writings of science and science fiction. First proposed in 1895 by H.G. Wells in his famous story *The Time Machine*, it appeared in science only in 1949 when the famous logician Kurt Gödel showed, completely unexpectedly, that Einstein's theory of general relativity – which accurately describes gravity and the dynamics of the whole Universe – allows time travel to occur. To Einstein's amazement, Gödel found exact solutions of his equations which described possible universes in which time travel was possible. These universes are quite unlike our own: they were *spinning* rather than expanding. People have speculated that Gödel found time travel an appealing feature of a universe because of his paranoid fears about death. He saw a means by which it might be possible for him to live on in some strange way.

Our experience of space and time is rather limited. We have been to the Moon. We have personal experience of just decades of years, have historical records of thousands of years, and can dig up fossil rocks that stretch back a few billion years. But all this experience is of parts of the

Universe where conditions are very mild: gravity is weak, densities are low, and large objects move much more slowly than light. This may not be a trustworthy guide to what is possible in more extreme circumstances. In Figure 11.1, we describe some of the different ways that have been found to build a time machine without violating the known laws of Nature.

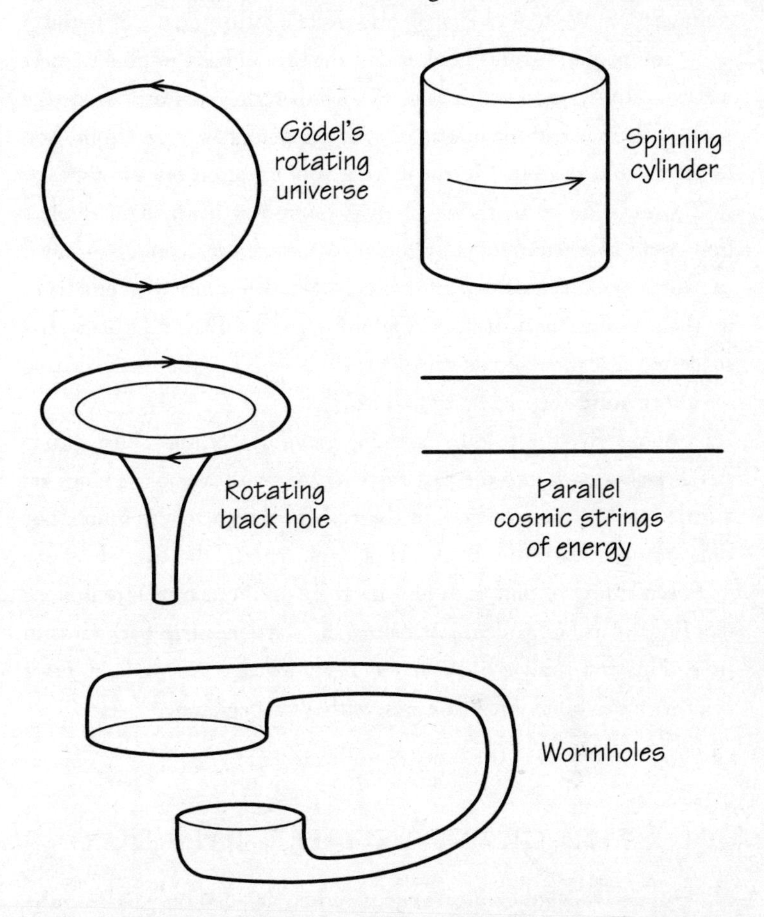

Fig 11.1 *A suite of different possible time travel devices that physicists have discovered could exist in principle, according to Einstein's general theory of relativity. Almost all exploit the presence of rotation in a universe to distort space and time to such an extent that closed paths in time become possible.*

If time travel can occur, then we seem to be facing inconsistency in Nature. It looks as if we could create factual contradictions by changing the past in ways that could not give rise to the present. You could bring about the death of your ancestors, so as to exclude the possibility of your own birth. Your current existence would then seem to constitute a logical contradiction. We could seemingly also create information out of nothing!

You could read this book today, then travel back in time to meet me as a student and tell me all the words that you learnt from the book. Where would the information in the book have come from? You learnt it from me; but I learnt it from you! Creation out of nothing.

Clearly, the entire theory of the evolution of life by natural selection could be circumvented by this means: organisms could be trained to avoid, or be forewarned of, hazards that they must overcome later in their evolutionary history. Oxford physicist David Deutsch has suggested that there should exist a principle which prohibits the getting of information for free by time travel.[23]

Books give rise to other seeming paradoxes. Suppose that a time traveller goes back into the past carrying a copy of a book. Things are a little busy in the past and, in the rush to return to the future, our time traveller leaves the book behind, underneath the tree where she had been sitting reading it. It remains there until our time traveller one day finds it in her garden, just before she starts her trip backwards in time. This is a strange book. It was never written, never edited, never printed, never bound: it just exists, without a beginning.[24]

THE GRANDMOTHER PARADOX

'so they go on in strange paradox, decided only to be undecided, resolved to be irresolute, adamant for drift, solid for fluidity, all-powerful to be impotent.'

Winston Churchill[25]

Logical paradoxes of the 'what-if-I-killed-my-grandmother' type constitute a genre called 'Grandmother Paradoxes' by philosophers interested in time travel. They appear to beset any form of backward-in-time travel (as opposed to forward-in-time travel). It has been a prominent component of science fiction stories about time travel ever since the scenario of machine-borne time travel began with Wells's classic.

Some regard the Grandmother and Information-out-of-nothing paradoxes as a proof that time travel is forbidden in our Universe. (A weaker version of this prohibition would allow time travel only so far as it did not create changes in the past.) For example, the well-known science fiction writer, Larry Niven, wrote an essay in 1971 entitled *The Theory and Practice of Time Travel* in which he enunciated 'Niven's Law' of time travel:

> 'If the Universe of Discourse permits the possibility of time travel and of changing the past, then no time machine will be invented in that Universe.'

Niven is convinced that time travel is equivalent to the introduction of irreconcilable inconsistency in the Universe and must be prohibited by some consistency principle deep within Nature's laws.

Nor are such worries confined to science fiction writers. In 1992, the physicist Stephen Hawking gave the same general 'no time travel' idea a name: the *Chronology Protection Conjecture*.[26] Hawking believes that time travel into the past cannot be possible because 'we have not been invaded by hordes of tourists from the future' arriving to watch or change great moments in history. But we might well ask how we would know what to look for, or how we would tell whether the 'normal' course of history was being disrupted by time travellers: maybe John F. Kennedy would have started World War III in 1965 if he had not been assassinated a year earlier. Hawking's Chronology Protection Conjecture asserts that the laws of physics prevent the creation of a time machine, except at the beginning of time, if such there was, when there is no past to travel into.

CONSISTENT HISTORIES

'Too much consistency is as bad for the mind as it is for the body. Consistency is contrary to nature, contrary to life. The only completely consistent people are the dead.'

Aldous Huxley[27]

Another response to the time travel paradoxes is to allow time travel *to occur so long as it does not produce logical or physical paradoxes* – for example, it must not create information or energy out of nothing. Here is an example of a consistent history approach to a time travel paradox:

Imagine that you travel back in time and prepare to shoot yourself when you were a baby. You are determined to create a paradox of fact in the Universe. You take aim at yourself when you are being held in your mother's arms. You move to pull the trigger, but an old injury to your shoulder, caused by your mother dropping you when you were a baby, sends a spasm down your arm and causes your shot to miss its target. But the sound of the gunshot is enough to startle your mother, who drops the baby on the ground, injuring his shoulder. Consistent histories make time travel safe for historians.

TOURISTS FROM THE FUTURE

'It has been said that though God cannot alter the past, historians can; it is perhaps because they can be useful to Him in this respect that He tolerates their existence.'

Samuel Butler[28]

This 'tourists from the future' problem has a long history. It is known as the 'cumulative audience paradox' amongst science fiction writers, after Robert Silverberg's explicit introduction of it in 1969.[29] As time travellers flock to the past, the worry is that an ever-increasing number of people accumulate at significant events in our history. Silverberg argues that events like the Crucifixion would attract billions of time travellers, yet 'no such hordes were present' at the original event. More generally, we will find our present and past increasingly clogged with voyeurs from the future:

> 'A time is coming [when time travellers] will throng the past to the choking point. We will fill our yesterdays with ourselves and crowd out our own ancestors.'

These visitors would, in effect, be gods: they would have control over time and access to all knowledge. Maybe the level of technical knowledge that makes such travel possible also reveals the deep problems that its exploitation would create, and wisdom ensures that the knowledge is never exploited. It offers the possibility of destroying the coherence of the Universe in the same way that our knowledge of nuclear physics offers us the means of destroying the Earth. In fact, American writer John Varley, in his science fiction story *Millennium* (1983), is worried that

> 'Time travel is so dangerous it makes H-bombs seem perfectly safe gifts for children and imbeciles. I mean, what's the worst that can happen with a nuclear weapon? A few million people die: trivial. With time travel we can destroy the whole Universe, or so the theory goes.'[30]

This type of 'where are they' argument against time travellers from the future is rather reminiscent of Enrico Fermi's famous 'Where are they?' response to claims that the Universe should be over run with advanced extraterrestrials.[31] Of course, we see no evidence for them at all. Some possible reasons for the absence of advanced extraterrestrials are the following 'magnificent seven':

1. There aren't any yet able to signal. We are the most advanced life-form within communicating range.

2. Technological civilisations cannot survive for long enough to become super-advanced. They blow themselves to bits, get wiped out by asteroidal impacts, or succumb to internal problems – disease, exhaustion of raw materials, or irreversible degeneration of their environment by pollution.

3. There are so many civilisations, and ours is a fairly average example of which there are millions of others. Therefore the most advanced extraterrestrials have no reason to take any special interest in us. We are just like another species of common insect.

4. Advanced extraterrestrials have a rigid code of non-interference in the histories of more primitive civilisations. We are like a cosmic game reserve; we are being studied but in a non-intrusive fashion.

5. Advanced extraterrestrials exist, but communicate only with technology at levels exceeding our own. In this way they require that a particular level of scientific maturity is required before any civilisation can join the 'club'.

6. Time travel is possible but extremely improbable. It requires time-travelling paths to lead to logically consistent histories. This requirement is so restrictive that time travel never leads to observable consequences in practice, except in the realm of elementary particle physics.

7. Advanced extraterrestrials exist but they are nano-sized, close to the size of atoms and molecules. Our advanced technologies are becoming smaller and more energy efficient, and civilisations that are millions of years ahead of us have become imperceptibly small to our astronomical probes.

Each of these responses to why we don't see evidence of *space travellers* can be recast as an explanation for why we see no *time travellers.* But in the time travel case, there is the possibility of a fundamental self-prohibition being imposed by super-advanced extraterrestrials, because they understand more fully that there would be grave consequences for the coherence of the whole of space-time if time travel were indulged in. Space travel may be expensive or impractical, but it is not potentially damaging to the fabric of reality.

TIME TRAVELLERS IN THE FINANCIAL WORLD: PERPETUAL MONEY MACHINES

'Time goes, you say? Ah no!
Alas, Time stays, we go.'

Austin Dobson[32]

The most novel version of the 'Where are they?' argument must surely be that proposed by the Californian economist Marc Reinganum, who wrote an article with the title 'Is Time Travel Possible?: A Financial Proof'.[33] He argues that the fact that we see positive interest rates proves that time travellers do not exist (he also claims that it means they cannot exist, which does not follow at all). The reasoning is simply that time travellers could use their knowledge to make such huge profits all over the investments and futures markets that interest rates would be driven to zero. They are not zero, therefore time-travelling investors do not exist!

In Douglas Adams's *The Restaurant at the End of the Universe* a similar scenario is conjured up, from which the book takes its name.[34] There is a restaurant located far in the future – at the very end of time. Diners are taken by time machine to the restaurant where they eat and drink extravagantly while watching the final destruction of the Universe through the restaurant's windows. At the last moment there, dinner ends and they are whisked back home by time machine. The bill for the evening of entertainment is stupendous, but almost anyone can afford it simply by depositing a single penny in an interest-bearing account in their own time, such is the interest that amasses by the end of time.

Alas, the time travellers' dining club doesn't make sound economic sense. If time travel were to become possible, then a time traveller from the year 3001 could carry \$1 back to the year 2001. Suppose the interest rate was 4%. When he got back home and checked out the account he would find that the \$1 had grown by compound interest to a staggering

$$\$1 \times (1 + 0.04)^{1000} = 108 \text{ million billion dollars !!}$$

– enough to buy quite a lot of planets at today's prices! If this isn't enough for his needs in 3001, he can just take some of it back to 2001 and invest for the future again. Clearly, if time travel becomes routinely possible at no cost to the traveller, then the interest rates through history would need to be 0% or time travellers could use the banking and investment system as a perpetual money machine.

Notice that negative rates of return are also inconsistent with no-cost time travel. Suppose an investment is worth \$1 at first and then falls to 50 cents in value subsequently. Again, time travellers could build a money machine. They could short-sell their investment in the first period (when it is worth $1), teleport in time to the epoch when it is worth 50 cents and repurchase it. Alternatively, they could just buy it when it is worth 50 cents and travel back in time to sell when it is worth $1. Either way, time travellers earn a profit of 50 cents. Again, these profits only disappear when the interest rates are zero. Someone once said that Einstein proved that time is money. Perhaps this is what he meant.

Notice that even if the technology to build time machines did not exist until thousands of years in the future, we should still observe zero interest rates today.

This argument reminds me of one that can be used against claims that clairvoyance or other psychic powers would give their possessors the ability to get rich from any form of gambling. Why bother bending spoons and guessing cards when you can win the National Lottery every week? If these powers existed in humans, then they would have bestowed such advantages upon their recipients that they would have become dominant in many ways, and the ability should have evolved throughout successful human sub-populations.

There might, of course, be rather more mundane restrictions which simply render time travel uneconomic. Tourists from the future might require such enormous energy expenditure, that the whole idea is always hopelessly impractical, even if it is possible in principle. Or maybe the companies that run the transport systems are always having to stop the service because of 'operational delays' or over-zealous safety restrictions and it never happens.

WHY YOU CAN'T CHANGE THE PAST

'You must remember this,
A kiss is still a kiss,
A sigh is just a sigh;
The fundamental things apply,
As time goes by.'

Herman Hupfeld[35]

Should we be persuaded by these 'changing-the-past' arguments against the possibility of time travel? There is something not quite coherent about all these arguments about *changing* the past. The past was what it was. You cannot alter it and expect the experienced present to still exist. We might have been there influencing it; but how could there be two pasts – one which was, and another which would have been if we had intervened, but which are in some way inconsistent with one another? If you could travel back in time to prevent your birth, then you would not be here to travel backwards in time for that purpose.

If we look more closely at the logic of the Grandmother Paradoxes, we see that there is a nagging worry about their coherence. Time travel must not involve undoing or changing the past in a manner that implies that there are two pasts: one without your intervention and one with it. If you travel back to influence some historical event, then you would have been part of that event when it occurred. A contemporary historical record would have included your presence (if you were noticeable). Time travellers do not change the past, because they cannot do anything in the year 1066 that was not actually done in 1066. Someone can be present at an event in the past and contribute to the record of what happened in history; but, that is quite different from the presumption that they can change the past. The past can be affected but not changed. If a change occurs, we can ask for the date when that change occurred. In the same way, the American philosopher Larry Dwyer has argued,

> 'Time travel, entailing as it does backward causation, does not involve changing the past. The time traveller does not undo what has been done or do what had not been done, since his visit to an earlier time does not change the truth value of any propositions concerning the events of that period . . . It seems to me that there is a clear distinction to be made here, between the case where a person is presumed to change the past, which indeed involves a contradiction, and the latter case where a person is presumed to affect the past by dint of his very presence in that period.'[36]

Usually we think of the passage of time as linear. Time travel is equivalent to this line closing up into a circle. Imagine a straight line of people walking one behind the other. There is a clear notion of who is behind and who is ahead of somebody else (Figure 11.2). This is like linear time. You can always say whether an event lies in your future or your past.

But now suppose the line of people are walking around in a circle. Then *locally* (just looking one place ahead or behind you) it appears clear that somebody is ahead or behind you. But overall the idea of ahead or behind does not have any meaning when you think about the whole circle – any one person is ahead of everyone and behind everyone. It can no longer be said that anyone is ahead or behind anyone else. They are both. Either the entire circular history is self-consistent or it is not.[37] And so it is with a time-travelling history. There is no unambiguous notion of past and future. You can't 'change' the past. There is just a logically consistent sequence of events along a closed loop of time. It is what it

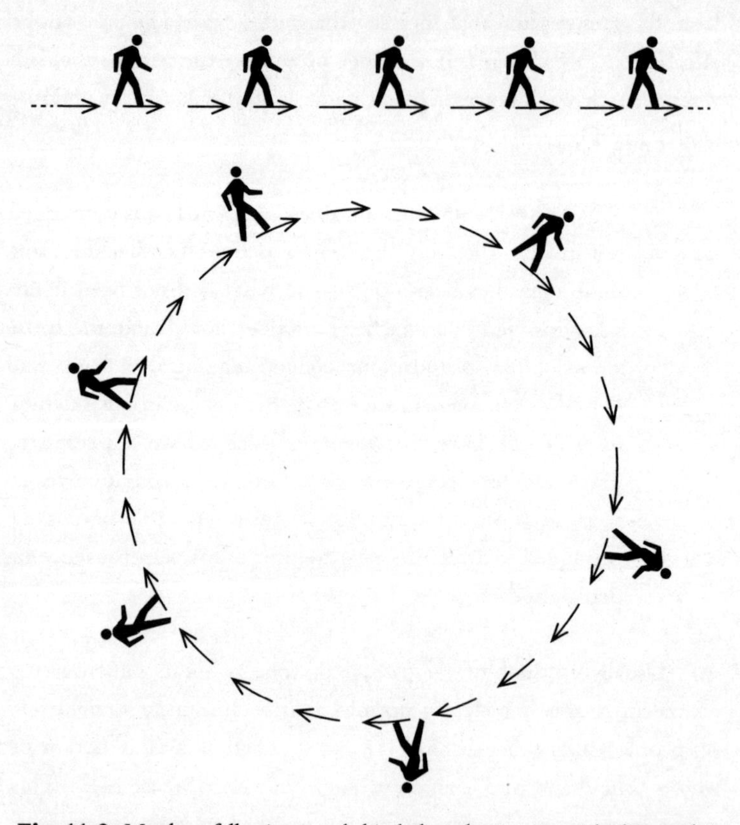

Fig 11.2 *Marchers following one behind the other in a straight line and in a circle. When they march in a circle everyone is both in front and behind everyone else. When they walk in a straight line everyone is either in front or behind everyone else.*

is and it was what it was. You can be part of the past, but you can't change it. Your experience will periodically recur as, in Nietzsche's words, 'the eternal sand-glass of existence will be turned once more'.[38]

So, the message of our trip through time is that time travel does not provide the secret of living forever. We don't know if it is a practical possibility. All the examples that have been found in Einstein's equations are weirdly extreme and bear little relation to situations that we have seen in the Universe. But who knows? Maybe we haven't been looking in the right places. Closed paths in time offer a way to have an endless future of only finitely many possibilities.

INFINITY – WHERE WILL IT ALL END?

'I think we agree, the past is over.'

George Bush[39]

Our revels now are ended. We have dug just a little in the rich seam of ideas that runs through the history of human thinking about the large, the small, and the in-between. The clash of the infinite with the finite is a dilemma that is deeply imbedded in our minds. Wherever we look, we find its manifestation in our thinking about the Universe, about counting, about the continuation of our consciousness here or in some other realm, and about where we have come from. We are drawn to the limits of time, space and matter in our search for answers to the ultimate questions about the Universe. There we find infinities of all shapes – and even sizes – and learn that it is unwise to treat them all the same. Unlike the ancients, we do not exclude them from our conceptions of the world, but nor do we always welcome them.

We have learnt that infinity is a player of great significance who appears on the stage only when the crucial questions of existence are raised. Infinity offers its services when we seek to know if the Universe began or whether it will ever end, whether life will always be part of its landscape, and whether there are tasks which can never be accomplished. Infinity challenges us to contemplate the duplication of ourselves and all that we hold dear, and to ponder the cogency of all possibilities, potential and actual. It undermines our sense of the precious by suggesting a randomly infinite universe will eventually conjure up the works of Shakespeare, somewhere, as if created by a regiment of monkeys armed with typewriters.[40]

Infinity also seeks to guard us from taking the wrong path in our quest to unravel the deepest of Nature's secrets about the ultimate structure of mass and energy. Once infinity seemed like an evil spirit that was determined to confuse our way, but we have come to see that it is a surer guide than ever we thought to the true path. Take one step off that path towards the Theory of Everything and infinite alarm bells ring. Our hope is that there is only one finitely specified path. It will lead to places where experiments may not reach, where observations cannot penetrate. The quest to understand the nature of matter and the Universe of space and time may come to rely uniquely and completely upon the beckoning or the avoidance of the infinite. We will need to know it better than we know ourselves.

∞

∞

∞

∞

∞

∞

∞

∞

∞

∞

∞

∞

∞

∞

∞

∞

∞

∞

∞

Notes

'Theories, theories, myriads upon myriads of them, streamed over me like windborne leaves, like the contents of some titanic paper-factory flung aloft by the storm, like dust-clouds in the hurricane advance of the mind. Gasping in this vast whirling aridity, I almost forgot that in every mote of it lay some few spores of the organic truth, most often parched and dead but sometimes living, pregnant, significant.'

Olaf Stapledon, *Last and First Men*

'her mind seemed to work independently of her precious library, but at the same time she depended for inspiration on the presence of her books, a silent living presence whose company sustained and reassured her.'

John Bayley, of Iris Murdoch

preface

1. Interview with Stephanie Merritt, 'Move over Coetzee', *Observer* Review, 28 September 2003, p. 16.
2. Quoted by Van Gogh's biographer, René Huyghe, in *Van Gogh*, Screpel, Paris, 1972.

chapter one

Much Ado about Everything

1. A. Lerner, title song of the 1965 musical *On a Clear Day.*
2. M. de Unamuno, *Tragic Sense of Life,* trans. J.E. Crawford Flitch, Dover, New York, 1954, first pub. 1921.
3. Quoted in M. Dibden, *Medusa,* Faber, London, 2003, p. 240.
4. http://www-gap.dcs.st-and.ac.uk/~history/Curves/Lemniscate.html
5. T. Stoppard, *Rosencrantz and Guildenstern Are Dead,* Act 2, Faber, London 1967.
6. Caius Glenn Atkins, *Greatest Thoughts on Immortality,* quoted in *The Pan Dictionary of Religious Quotations,* ed. M. Pepper, Pan, London, 1989, p. 250.
7. http://stingetc.com/lyrics/windmills.shtml
8. Augustine, *De Trinitate.*
9. This is a more subtle matter than may at first appear. The appearance of the sky varies significantly with latitude on the Earth's surface, and the myths and legends that were inspired by the appearance of the sky vary in systematic ways that reflect features like the apparent rotation of the night sky around the celestial poles. For a fuller account of these influences see J.D. Barrow, *The Artful Universe,* Oxford University Press and Penguin, 1995, and references therein.
10. This appears to be a pun on Babel, meaning the 'gate of God' in Sumerian and Babylonian, and *balal* meaning confusion.
11. Genesis 11:4.
12. Quoted in E. Maor, *To Infinity and Beyond: a cultural history of the infinite,* Princeton University Press, 1991, p. 138.
13. For a discussion see J.D. Barrow, *Pi in the Sky: counting, thinking, and being,* Oxford University Press and Penguin, 1992.
14. Robert Pirsig, *Lila,* Bantam, New York, 1992.
15. W. Blake, *Auguries of Innocence,* Dover, New York, 1968.

16. Photographs of Islamic tilings by Patrick Syder; copyright © Patrick Syder Images.
17. Reproduced by permission of Roger Penrose.
18. Archimedes brilliantly used this observation to produce a very accurate calculated approximation to the value of π between 3.14282 and 3.14090, by approximating the circle by a 96-sided polygon.
19. For the details see the fuller discussion in J.D. Barrow, *The Constants of Nature*, Jonathan Cape, London, 2002, pp. 117–18.
20. The psychologists George Lakoff and Rafael Núnez, in their book *Where Mathematics Comes From* (Basic Books, New York, 2000, chap. 8) have argued that there exists a 'Basic Metaphor of Infinity' (BMI) which is part of our conceptual system. They argue that there are several mathematical structures that appear similar from a cognitive point of view. However, they interpret this to mean that the brain's conceptual apparatus is not a harmless vehicle by which we apprehend the nature of reality, but a processor that completely shapes the mathematics that we use by adapting the way we deal with finite things on an everyday basis. The discussion we have given in this chapter can be seen as showing some of the ways in which the concept of infinity enters our minds. However, we would argue that it has an existence independent of that of our minds. Just because our minds shape our apprehension of it in unavoidable ways does not mean that there is nothing to the concept of infinity except that injected by our minds.
21. The third and fourth paradoxes are of less interest. Zeno's Third Paradox considers an arrow in flight. At any moment of time the arrow is at a particular fixed place. Hence, suggests Zeno, it cannot be moving. Zeno's Fourth Paradox asks us to consider three rows of people:

XXXX
YYYY
ZZZZZ

The first row, XXXX, is stationary; the second row, YYYYY, is moving to the right at maximum speed and the third row, ZZZZZ, is moving to the left at maximum speed. But, says Zeno, relative to each other the Ys and Zs are going at *twice* the maximum possible speed, which is impossible, so motion must be impossible. Neither of these paradoxes is especially interesting given our modern understanding of motion. For example, special relativity does maintain that there is a maximum speed, c, equal to the speed of light in a vacuum. But if Y moves with speed u relative to X and Z moves with speed v relative to X, the relative speed of Z with respect to Y is not given by $u + v$, as Zeno assumed, but $(u + v)/(1 + uv/c^2)$, and we notice that this never exceeds c. When $u = v = c$, the relative velocity between Y and Z still equals c, as Einstein and others showed 2,500 years later.

chapter two

Infinity, Almost and Actual, Fictitious and Factual

1. B. Pascal, *Pensées*, ed. A. Krailsheimer, Penguin, London, 1966, fragment 418.
2. H. Weyl, *God and the Universe: the Open World*, Yale University Press, New Haven, 1932.
3. It is interesting to consider why the mathematical theory of chance and probability took so long to emerge. Games of chance existed in many ancient cultures, and the same cultures developed other branches of mathematics and science. One possibility is that early gaming devices were asymmetrical in shape, usually made of knuckle bones (*stochos* – from which we take the word stochastic, meaning random – is the Greek for aim). This meant that each one of these devices was unique and there was no general theory based upon equally likely outcomes as there is with a symmetrical playing die. The other possibility is that chance was regarded as the way in which the gods spoke and influenced the world. This

can be seen in the Old Testament of the Hebrew Bible where the casting of lots was used to determine the Divine will. A well-known example is the decision to cast the prophet Jonah overboard. The High Priest also had a way to determine whether God was saying Yes or No to his requests by drawing two-sided plates from his garment. In such a climate of belief it would be very unwise, and certainly blasphemous, to start systematically studying the ways of chance.

4. J.D. Barrow, *The Book of Nothing*, Jonathan Cape and Vintage, London, 2000.
5. Aristotle, *Physics*.
6. Ibid., III 206a, 14–25.
7. Ibid., 206b, 33–207a, 15.
8. J. Lear, 'Aristotelian Infinity', *Proc. Aristotelian Soc.* 80, 187–210 (1979).
9. See Aristotle, *Posterior Analytics*. I.22; Lear, op. cit., p. 202.
10. In *Physics* IV 14, 223a, 16–28 Aristotle writes: 'It is worth investigating how time is related to the soul and why time seems to be in everything both in earth, in sea and in heaven . . . Someone might well ask whether time would exist or not if there were no soul. If it is impossible for there to be someone who counts, then it is impossible that something be countable, so it is evident that neither is there number, for number is either the thing counted or the countable. If there is nothing capable of counting, either soul or the mind of soul, it is impossible that time should exist, with the soul not existing, but only the substratum of time, if it is possible for change to exist within soul.'
11. The hymn writer John Newton appears to have had a strong interest in mathematics in his early life. We learn that when he was involved in the slave trade, he experienced a period of house arrest on board ship and in Gambia, after being falsely accused of stealing. 'He was at this point so destitute that if a boat came to visit the island he would hide for shame of his condition. One thing, oddly enough,

that he managed to keep was a volume of [Isaac] Barrow's Euclid; a study in mathematics. He would go to a remote corner of the island and work out mathematical diagrams in the sand. He said, "Thus I often beguiled my sorrows . . .'" but he relinquished this mathematical interest upon his conversion, feeling it to be a useless treasure; see http://www.cyberhymnal.org/htm/a/m/amazgrac.htm and http://www.geocities.com/Heartland/Pointe/4495/biography.html. The famous hymn *Amazing Grace* was written by Newton whilst vicar at Olney, and was published in *Olney Hymns*, by W. Oliver, London in 1779. However, the remarkable last stanza quoted here is an addition to Newton's original that appears first in the text of Harriet Beecher Stowe's anti-slavery novel *Uncle Tom's Cabin*, published in 1852. The verse appears in the text of chapter 38 when Tom has a vision as he lays dejected by the fire and hears the words of Newton's hymn but it is recorded for the first time with this extra stanza that has been imported from another hymn, 'Jerusalem My Happy Home' which appeared in *A Collection of Sacred Ballads* published in Virginia in 1790. Moving stanzas between hymns (so called 'wandering stanzas') was common practice amongst the slave community. The imported verse looks out of place in Newton's hymn because the 'I' and 'me' of the earlier verse suddenly changes to 'we'.

12. Augustine, *City of God*, Book XII, chapter 18.
13. A simple example of this sort is the transformation that changes the number x into tanh(x).
14. Pascal, *Pensées*, op. cit.
15. Engraving of Blaise Pascal reproduced from Léon Brunschvicg, *Pascal*, Éditions Rieder, Paris, 1932.
16. This predates the limits on signalling velocity introduced by the theory of relativity. Even later, in Newton's time, the action of gravity was supposed to occur instantaneously, and hence at infinite speed.

17. B. Pascal, *De l'esprit géométrique* (1657–8), in *Great Shorter Works of Pascal,* trans. E. Caillet and J. Blankenagel, Westminster Press, Philadelphia, 1948, p. 195.
18. Ibid., p. 196.
19. Pascal, *Pensées,* op. cit., p. 91.
20. These are the so called Planck units of length and time defined by combinations of the speed of light, Newton's constant of gravitaion, and Planck's quantum constant. Numerically, this minimum length is approximately 10^{-33} cm and the minimum time is approximately 10^{-43} seconds. If one tries to explore the structure of space and time on dimensions smaller than these, then their very nature undergoes a quantum gravitational transformation whose nature is not yet understood. For further discussion of these fundamental units see J.D. Barrow, *The Constants of Nature,* Jonathan Cape, London, 2002.
21. Pascal, *Pensées,* op. cit., fragment 199.
22. Galileo Galilei, *Two New Sciences,* trans. S. Drake, University of Wisconsin Press, Madison, 1974, p. 34.
23. R. Descartes, *Principles of Philosophy,* 26, quoted in M. Blay, *Reasoning with the Infinite,* University of Chicago Press, 1993, p. 9.
24. Descartes, op. cit., 26, quoted in Blay.
25. Ibid., 27, quoted in Blay.
26. For further discussion and references see J.D. Barrow and F.J. Tipler, *The Anthropic Cosmological Principle,* Oxford University Press, 1986, chap. 2, n. 245.
27. We do not (and neither do theologians it seems) dwell on the distinctions that are possible regarding the size of infinite sets.
28. N. Cusa, *On Learned Ignorance,* trans. J. Hopkins, Banning Press, Minneapolis, 1985, original pub. 1444.
29. This is essentially the argument of Popper and MacKay about the logical impossibility of predicting a person's future actions if the prediction is made known to them; see J.D. Barrow, *Impossibility,* Oxford University Press, 1998.

30. There have been interesting additions to this idea as a result of the theory of evolution. Our senses have evolved as a result of selective pressure imposed by the nature of true reality. Thus our eyes and ears have a form and structure that tell us something about the true nature of light and sound; for further discussion, caveats, and references see J.D. Barrow, *The Artful Universe*, Oxford University Press and Penguin, 1995.
31. M. Kline, *Mathematical Thought From Ancient to Modern Times*, Oxford University Press, 1972, p. 994.

chapter three

Welcome to the Hotel Infinity

1. F. Morgan, *The Math Chat Book*, Math Assocn. America, 2000, p. x.
2. G. Hoffnung, Speech at the Oxford Union, 4 December 1958, quoting from a supposed hotel brochure supplied by an Austrian landlord.
3. Pictures from the play *Infinities* (Fig. 3.1, 3.2, 5.2 and 8.1) reproduced by kind permission of Serafino Amato.
4. David Hilbert was one of the world's foremost mathematicians in the first part of the twentieth century. The story of the infinite hotel seems to have been attributed to him, but he never wrote about it. It became better known when briefly described by George Gamow in his book *One Two Three . . . Infinity: facts and speculations in science*, Viking, New York, 1961 (first pub. 1947), pp. 17–18. Gamow acknowledges 'the unpublished, and even never written, but widely circulating volume: "The Complete Collection of Hilbert Stories" by R. Courant'. Courant was one of Hilbert's closest collaborators. Hilbert's views about mathematical infinity were summarised in his article 'On the Infinite' which is reprinted in P. Benacerraf and H. Putnam, *Philosophy of Mathematics: selected readings*, 2nd edn, Cambridge University Press, 1983, pp. 183–201.

5. J. Cornwell, *Hitler's Scientists,* Viking, London, 2003, p. 120.
6. N. Ya. Vilenkin, *In Search of Infinity*, Birkhäuser, Boston, 1995, p. 43.
7. The origin of this interesting phrase, which I always thought as a child had something to do with the board game of snakes and ladders, derives from the early days of radio in Britain when the first commentaries were given on football matches. A numbered grid of the pitch was published in the *Radio Times* and the commentator could tell the listener where the ball was by giving the numbered square. If a team's attack broke down so that they were forced to play the ball back to their own goalkeeper, who was located on square 1, then they were described as being 'back to square one'.
8. Prime numbers cannot be divided exactly by any whole number other than themselves and one.
9. The maths student reappears to show him that if a guest occupied room n in hotel m, then if $n \geq m$ that guest will be put in room $(n-1)^2+m$, and if $n \leq m$ they will be put in room $m^2 - n+1$.
10. S. Leacock, 'Boarding House Geometry', in *The Best of Stephen Leacock,* vol. 1, ed. J.B. Priestley, Humorbooks, Sydney, 1966, p. 26.
11. P.E.B. Jourdain, *The Philosophy of Mr B*rtr*nd R*ss*ll,* Allen and Unwin, London, 1919, p. 66.
12. John Cage wanted to create a piece of music that was the musical analogue of absolute zero of temperature (minus 273 degrees centigrade). It is called *4′ 33″*. It was originally performed with a pianist in formal evening dress sitting motionless in front of a grand piano for 4 minutes and 33 seconds (which equals 273 seconds).
13. J.D. Barrow, *The Book of Nothing,* Jonathan Cape and Vintage, London, 2000.

chapter four

Infinity Is Not A Big Number

1. Vedic mantra (3rd or 2nd millennium BC) explaining the Indian understanding of the abstract notion of infinity thus (in Sanskrit):

'purnamadah purnamidam purnat purnamudachayte purnasya purnamadaya purnamevavashishyate'. I am grateful to Subhash Kak for this quotation and its translation.

2. Television news conference, www.geocities.com/Ama 51/Zone/7474/blquayle.html
3. A 'sophisma' should not be confused with a 'sophism'. The latter is a rather pejorative term used by the ancients about empty philosophising. Sophisma is a technical term in medieval logic that describes particular statements that are set up for critical analysis. They are not questions or problems but ambiguities or puzzles which require interpretation or which challenge a logical system. A modern genre that is comparable would be self-referential statements.
4. www.wow4u.com/questioning.
5. For a collection of information and pictures related to Galileo and his work see http://es.rice.edu/ES/humsoc/Galileo/
6. Extracted from *Dialogues Concerning Two New Sciences* (First Day sections 78 & 79).
7. Portrait of Albert of Saxony reproduced from www.gap.dcs.st-and.ac.uk/~history/BiogIndex.html
8. J. Royce, *The World and the Individual*, Macmillan, London, 1901, supplementary essay section III, Pt. I, pp. 504–5.
9. I Thessalonians 5:21.
10. If S(n) is the sum of the first n terms of the harmonic series, then Oresme's argument shows that $S(2^n) \geq (n+2)/2$ and it grows steadily to infinity as n increases.
11. Remarkably, the infinite series of alternating terms $1 - \frac{1}{2} + \frac{1}{3} - \frac{1}{4} + \frac{1}{5} - \frac{1}{6} +$ has a *finite* sum, equal to the natural logarithm of 2, or about 0.693.
12. This reasoning does not apply to sports records because they are not random occurrences.
13. R.M. Dickau, http://www.prairienet.org/~pops/BookStacking.html

14. Specifically, for N = 1 the overhang is 0.5, for N = 2 it is 0.75, for N = 3 it is 0.917, for N = 4 it is 1.042.
15. Irving Berlin, *Let's Face the Music and Dance.*
16. Besides the examples we have already given, it also tells the probability of finding bunches of 1, 2, 3, 4, 5, 6, 7, . . . n cars in a flow of traffic because the bunches are created by 'record' lowest speeds in the flow.
17. The Norwegian Academy has recently created an international prize for mathematics, to rival the Swedish Academy's long-standing Nobel Prizes in science, literature, economics and peace, called the Abel Prize in honour of Niels Abel (1802–29); it rhymes nicely with Nobel and has similar monetary value (approximately one million dollars). The first winner, in 2003, was the French mathematician J.P. Serre.
18. Photograph of Georg Cantor with his wife, Vally, *c.* 1880, reproduced from Joseph Warren Dauben, *Georg Cantor: His Mathematics and Philosophy of the Infinite,* Harvard University Press, 1979.
19. F. Hutcheson (1694–1746), *Inquiry into the Original of Our Ideas of Beauty and Virtue,* J. Darby, London, 1720, II, iii.
20. This proviso is to remove the ambiguity created by decimals that end with recurring 9's because 0.2699999 . . . , say, is the same as 0.2700000 . . . Hence by eliminating those decimal expansions that end in zeros we identify the number 27/100 by 0.26999 . . . and not by 0.27000 . . .
21. Portrait of Bernhard Bolzano reproduced from www.gap.dcs.st-and.ac.uk/~history/BiogIndex.html
22. Galileo used this analysis to 'show' that the circumference of any circle was equal in size to its central point! See, for example, the discussion of this by Bernard Bolzano in *Paradoxes of the Infinite,* Routledge & Kegan Paul, London, 1950.
23. A. Camus, *The Fall,* Vintage, New York, 1956, p. 45.
24. Cantor described a set as follows: 'A set is a Many that allows itself to be thought of as a One.'

25. Because there is an analogous nesting that uses the power set idea. If we have a thought, then we can have thoughts about that thought and so on. The information conveyed by the words 'the statement T is true' is different from that contained in the statement T itself.

Chapter five

The Madness of Georg Cantor

1. (1888–1973) US naturalist and environmental activist http://www.melodyonline.com/quotes2.html
2. A. Christie, *An Autobiography*, HarperCollins, London, 1998.
3. Quoted in E. Schechter, *Handbook of Analysis and its Foundations*, Academic, New York, 1998.
4. J. Dauben, *Georg Cantor*, Princeton University Press, 1990, p. 1.
5. Photograph of Leopold Kronecker, *c.* 1885, copyright © akg-images.
6. D. Burton, *History of Mathematics*, 3rd edn, Wm. C. Brown, Dubuque, IA, 1995, p. 593.
7. Dauben, *Georg Cantor*, p. 134.
8. Ibid.
9. Ibid., p. 135.
10. Ibid., p. 136.
11. Ibid., p. 147.
12. Letter 15 Feb 1896 to Esser, H. Meschkowski, *Arch. History of Exact Sciences*, 2, 503 (1965).
13. Dauben, *Cantor*, p. 147.
14. Unpublished letter to Carl Friedrich Heman 21 June 1888, Göttingen, Cod. Ms. Cantor 16, Nr. 83, S. 179.
15. II Chronicles 6:18.
16. Many years later finitism was relaunched as a radical crusade against mainstream mathematics by Luitzen Brouwer, a brilliant but unstable Dutch mathematician. Brouwer's campaign created a crisis within mathematics, this time with Brouwer – the proponent of finitism – as the victim; see J.D. Barrow, *Pi in the Sky*,

Oxford University Press and Penguin Books, London, 1992, chap. 5, for a detailed account.

chapter six

Infinity Comes in Three Flavours

1. Aristotle, *Physics.*
2. R. Eastaway and J. Wyndham, *Why do Buses Come in Threes?*, John Wiley, New York, 1999.
3. G. Cantor, *Gesammelte Abhandlungen*, p. 378; R. Rucker, *Infinity and The Mind*, Paladin, London, 1984, p. 9.
4. Cantor was unusual for his time in believing that the Universe was finite in age and in physical extent.
5. J. Dauben, *Georg Cantor*, Princeton University Press, 1990, p. 146.
6. Rucker, *Infinity*, p. 309.
7. A. Conan Doyle, 'The Boscombe Valley Mystery', *The Adventures of Sherlock Holmes*, Oxford University Press, 1993. This story was first published in the *Strand Magazine* 2, 401–16 in October 1891.
8. Detail from *The Thirty-six Views of Fuji*, by Katsushika Hokusai, reproduced by permission of the Metropolitan Museum of Art, New York.
9. In fact, the question of the infinite age of the Universe does not have a simple invariant interpretation. If there is no preferred scale of time, as Einstein's theory of general relativity requires, then one can choose ways of measuring time that are finite by one reckoning but infinite on another. For example, the change of time coordinate from t to log(t) results in the finite interval of t time from t = 0 to t = 1 changing into the infinite interval of log(t) time from minus infinity to 0. This 'problem of time' is one of the reasons why cosmologists like Hawking have sought to remove the explicit presence of 'time' from cosmological theories altogether, arguing that 'time' is not a fundamental defining feature of a quantum gravitational universe, but merely

an approximate concept that arises at low temperatures far from the apparent beginning of the expansion, see for example S.W. Hawking, *The Universe in a Nutshell*, Bantam, London, 2001.

10. J.D. Barrow, *Impossibility*, Oxford University Press, 1990.
11. See for example E.A. Milne, *Modern Cosmology and the Christian Idea of God*, Clarendon Press, Oxford, 1952 and E.T. Whittaker, *Space and Spirit: Theories of the Universe and the Arguments for the Existence of God*, Nelson & Sons, London, 1946.
12. S.W. Hawking and R. Penrose, *The Nature of Space and Time*, Princeton University Press, 1996.
13. A. Einstein and N. Rosen, *Physical Review* 48, 73 (1935).
14. P. Bergmann, in H. Woolf (ed.), *Some Strangeness in the Proportion*, Addison Wesley, MA, 1980, p. 156.
15. For many years, English fireworks bore a quaintly worded formal instruction about igniting them safely, that read 'Light the blue touchpaper and retire'.
16. The famous singularity theorems of Hawking and Penrose use a definition of singularity that is not necessarily characterised by the appearance of a physical singularity – merely the impossibility of extending the history of every possible path of a particle through space and time arbitrarily far in the past. If gravity is always attractive, then at least one of these possible histories must have a beginning. It is then a separate question to decide if all histories need to have such beginnings in our Universe, and whether the beginning is accompanied by a physical infinity, as in the familiar image of the beginning of an expanding Universe with infinite densities and temperatures; see J.D. Barrow and J. Silk, *The Left Hand of Creation*, Penguin, London, 1983 for a longer but elementary discussion.
17. The argument given is that if the singularities could be avoided, then the present expansion of the Universe could have been preceded by a collapsing phase. This would have led to a Universe with a succession of cycles of steadily increasing entropy (see R.

Penrose, in *The Nature of Space and Time*, Princeton University Press, 1996, p. 36. However, it is not certain that this would be a consequence of avoiding the initial singularity.

18. A. Ginsberg, *Howl*, 1956, p. 9.
19. The most general type of black hole can rotate and possess an electric charge as well as have mass. We shall discuss the non-rotating and uncharged black hole here. It is called the 'Schwarzschild black hole' after Karl Schwarzschild who, in 1916, found the solution of Einstein's equations that describes it. He did not know that his solution described the exotic situation that we now call a black hole. This term was invented by the American physicist, John A. Wheeler.
20. A preferred value for quantum effects to intervene and stop the collapse is 10^{96} gm per cc.
21. Of about 10 micrograms, the so called Planck mass.
22. There is no known reason why this black hole evaporation should not occur in principle. However, we do not know whether the very small black holes that would be needed if the effects were to be strongly visible today were formed in the early history of the Universe. Conditions may have been too smooth and quiescent.
23. G. Santayana, *The Unknowable*, in *Oxford Lectures on Philosophy 1910 to 1923*, Kessinger Whitefish, MT, 1924, p. 4.
24. R.D. Lunginbill, *Theology: the study of God*, http://associate.com/ministry_files/mirrors/ichthys.com/1Theo.htm
25. M. Wiles and M. Santer, *Documents in Early Christian Thought*, Cambridge University Press, 1977, p. 6, from Clement of Alexandria, *Misc.* 5, XII, 78–82.
26. This theological system, known as the apophatic way, from the Greek *apophatikos* – meaning 'negative' – is emphasised in Eastern Christianity and follows the interpretation of Gregory of Nyssa, 'Therefore St John, who also entered into this bright mist, says that no one has seen God at any time, meaning by this negation that the knowledge of the divine nature is

impossible not only for humans but for every created intellect', Gregory of Nyssa, *Opera Omnia*, J.-P. Migne, ed., Paris, 1863, p. 376d and A. Meredith, *Gregory of Nyssa*, Routledge, London, 1999.

27. See for example the discussions by A. Flew, *God and Philosophy*, Harcourt Brace, New York, 1966, and W.L. Rowe, *The Cosmological Argument*, Princeton University Press, 1975.
28. http://www.stats.uwaterloo.ca/~cgsmall/ontology.html
29. There is a string of formal deductions that can be made along these ontological lines. For example, that God exists. Assume the property of existence is an attribute of perfection. Define a *god* as any being that has all perfections. Then it follows that *All gods exist, that is, every god has the property of existence.* Now assume that given any perfection Q, if all things having Property Q also have the property of existence, then there is at least one entity having the Property Q. Then we can prove that *something exists,* that is, there is at least one entity that has the property of existence. Now, assume first, that given any collection of perfections, the property of having all the perfections in the collection is again a perfection, and second, there is a collection of perfections that contains all perfections. Then we can prove that *the property of being a god is a perfection and there exists at least one god.* Now assume that for any god g, the property of being identical to g is a perfection, then it follows that there is only one god. A collection of analogous results can be proved about the Devil if it is assumed that non-existence is an imperfection, and that given any imperfection A, if there is no existent entity having property A, then there is no entity at all having property A and the Devil possesses all imperfections. Hence, the Devil does not exist. See R. Smullyan, *5000 B. C. and Other Philosophical Fantasies*, St Martin's Press, New York, 1983 and also http://cs.wwc.edu/~aabyan/Philosophy/Ontology.html
30. J.D. Barrow, *Pi in the Sky*, Oxford University Press, 1992, p. 123.

C.A. Pickover, *The Paradox of God and the Science of Omniscience,* St Martin's Press, New York, 2002, p. 137.

31. F. Nietzsche, *Jenseits von Gut und Böse,* stanza 146.

chapter seven

Is the Universe Infinite?

1. Quoted in N. Rose, *Mathematical Maxims and Minims,* Raleigh, WC, 1988.
2. *Observer* colour magazine, London, 22 December 2002, p. 32.
3. The other way in which the infinite impinged upon ancient thinking about the Universe was in respect of its lifetime: did it have a beginning and will it have an end? Here the guiding intuition was not drawn from geography but from the finiteness of the human lifetime, tempered by possible belief in reincarnation and a cyclicity to all things which mirrored the cycles of the seasons, of growth and germination.
4. This work appeared as an appendix to a later edition of the work of meteorology first published by his father Leonard Digges under the title *A Prognostication* in 1553. Both were anxious to show that the study of astronomy was neither 'useless' nor 'impious'.
5. Uranus was discovered by William Herschel in 1781 and was originally named *Georgium Sidus* after his patron King George III, but this was never popular. The name Uranus was given to it later by the German astronomer J.E. Bode. Neptune was then discovered in 1846 through the work of Adams, le Verrier and Galle. Finally, Pluto was discovered by Tombaugh and Lowell in 1930.
6. P. Usher, *Bull. Amer. Astron.* Soc 28, 1305 (1996) and 'Shakespeare's Cosmic World View', *Mercury* 26 (1), January–February 1997, 20–3.
7. *Hamlet,* II. ii. 264.
8. Quoted in E. Maor, *To Infinity and Beyond: a cultural history of the infinite,* Princeton University Press, 1987, p. 198.
9. Translations of G. Bruno, *De l'infinito universo et mondi* (1584) from

D.R. Danielson (ed.), *The Book of the Cosmos,* Perseus, New York, 2000, pp. 40–4.

10. 'O, dear what can the matter be', c. 1792 and #494 in 'The Scots Musical Museum', V, 1796.
11. Satellite photograph of the earth at night reproduced by permission of NASA (http://visibleearth.nasa.gov/).
12. The slight mass difference between the neutron and the proton ensures that, no matter how the Universe begins, when it cools to 10 billion degrees K there is about one neutron for every 7 protons. This is the crucial starting condition for the subsequent bout of nuclear reactions that quickly establish the abundances of hydrogen, deuterium, helium and lithium that are seen today.
13. The increasing density of matter increases the rate of reactions that destroy deuterium and helium-3. This reduces their abundance and the destruction leads to more helium-4. The total abundances of helium-3 and deuterium are far smaller than that of helium-4 so that, whereas the increased destruction creates a significant depletion in their abundances, the increase in the helium-4 abundance at their expense is relatively small.
14. M.C. Escher Foundation, *M.C. Escher: the official website,* http://www.mcescher.com
15. M. Livio, *The Golden Ratio,* Headline, London, 2002, p. 157. The gravestone does not survive. The epitaph is known from a drawing that was made of it.
16. In 1537, Nunes published two works about the geometry of such optimal courses, followed by a synthesis and expansion in 1566 in Latin, where he uses the word 'rumbo'. The word 'loxodrome' (Greek *loxos* = 'oblique', *dromos* = 'bearing') is a later 1624 Latinisation by Willebrord Snell, of Snell's law of optics, of the Dutch word *kromstrijk* meaning 'curved direction', used by Simon Stevin in his description of Nunes's work. A loxodrome extends from Pole to Pole, and winds infinitely often around them in a logarithmic

spiral path. A rhumb line between two points of the same longitude is an arc of a great circle, but a rhumb line deviates most from a great circle when the two points have the same latitude. See W.G.L. Randles, 'Pedro Nunes' discovery of the loxodromic curve (1537): how Portuguese sailors in the early sixteenth century, navigating with globes, had failed to solve the difficulties encountered with the plane chart', *Journal of Navigation* 50, 85–96 (1997).

17. E. R. Harrison, *Darkness at Night*, Harvard University Press, 1987.
18. This idea had been proposed by Karl Schwarzschild following the growth of interest in non-Euclidean geometries. He used a spherical geometry to model the known Universe at that time. However, the theory of gravity assumed was still Newton's and the geometry was fixed. It was not changed by the presence and motion of matter as in general relativity.
19. Quoted by Sir Edward Burne-Jones in a letter to Lady Horner, F. Metcalf, *The Penguin Dictionary of Modern Humorous Quotations*, Penguin, London, 1987, p. 172.
20. We are reminded that the inverse-square law of gravity is very special mathematically, yet that specialness leads to unique properties of gravity that are certainly essential for any living observers to be present in the universe.
21. Wilkinson Microwave Anisotropy Probe.
22. Song 1939 by E.Y. Harburg, music by Harold Arien.
23. Translation of E. Borel, *Space and Time* (1922), Dover, New York, 1960, p. 246.
24. Some years ago there was speculation that in a situation like this it might be possible for an underdense region to 'escape' the final collapse of the Universe to a singularity. In fact under very general circumstances this is not possible, see J.D. Barrow and F.J. Tipler, *Closed Universes: their evolution and final state*, Mon-Not-R. astm. Soc. 216, 395 (1985).
25. Version of original quote by the Danish poet, Piet Hein.
26. The *average* period of Halley's Comet is seventy-six years. It varies

due to gravitational perturbations by the planets and the boiling off of its constituent gases. Between 239 BC and AD 1986 its period has varied between 76.0 and 79.3 years. These extremes were on AD 451 and AD 1066.

27. Born 1904.
28. See E.R. Harrison, *Cosmology*, Cambridge University Press, 1981, p. 250.
29. After the German astronomer Heinrich Olbers (1758–1840), who in 1826 wrote an article entitled 'On the Transparency of Space' which posed the memorable question 'Why is the Sky Dark at Night?' The detailed expression of the paradox uses the fact that the apparent brightness of a star of fixed intrinsic brightness falls inversely as the square of the distance the star is away from us. If we consider the number of stars on all spherical shells centred upon ourselves and add up the apparent luminosity they contribute to what we observe, then the total will tend to infinity as we allow the shells to grow infinite in radius. The light we receive from each shell is constant, independent of the radius of the shell. Adding up the light from an infinite number of shells of increasing radius, we receive an infinite total brightness.
30. Quoted in E. Maor, *To Infinity and Beyond: a cultural history of the infinite*, Princeton University Press, 1987, p. 205.
31. Halley's point, and even the subsequent publication by Olbers, attracted no attention at the time. Astronomers were more interested in the host of other theoretical and observational advances that were taking place. The night-sky paradox was almost completely ignored until Hermann Bondi included it in his influential 1952 book *Cosmology* (Cambridge University Press). He drew attention to the fact that it was the most basic of all cosmological observations and interpreted the darkness of the night sky as a proof of the expansion of the Universe. The most exhaustive analyses of the history and the many erroneous attempts to

resolve the paradox have been made by Edward Harrison who has maintained a life-long interest in the story; see for example E.R. Harrison, *Darkness at Night,* Harvard University Press, 1987, and the second edition of E.R. Harrison, *Cosmology,* Cambridge University Press, 2000, chap. 24.

chapter eight

The Infinite Replication Paradox

1. C. Green, *The Human Evasion,* Hamish Hamilton, London, 1969, p. 12.
2. Justin Scott, *The Shipkiller,* quoted in A. Pérez-Reverte, *The Nautical Chart,* Picador, London, 2002, p. 56.
3. No matter how small the probability of an event occurring, so long as that number is non-zero when multiplied by infinity it will give infinity. This is the number of occurrences of that event that will be found in an infinite universe.
4. Quoted by W. Kaufmann in his translator's introduction to F. Nietzsche, *The Gay Science* (translation of *Die Fröhliche Wissenschaft*), Random House, New York, 1974, p. 16.
5. F. Nietzsche, *Complete works,* vol. IX, Foulis, Edinburgh, 1913, p. 430.
6. F. Nietzsche, *The Will to Power,* trans. W. Kaufmann and R. J. Hollingdale, Vintage, New York, 1968, stanza 1066.
7. H. Spencer, *First Principles,* 4th edn, Appleton, New York, 1896, p. 550.
8. The argument for the unpredictability of the evolutionary process seems to have been made by commentators like Stephen Jay Gould on the basis of a belief in the chaotic unpredictability of the evolutionary process. However, this betrays a misunderstanding of chaotic processes. Although individual trajectories are unpredictable if they are not precisely known at some time, the averaged behaviours of a chaotic system generally evolved very smoothly, predictably, and

unchaotically. This is sufficient to ensure the robustness of evolution in the large against many small perturbations.

9. G.S. Kirk and J.E. Raven, *The Presocratic Philosophers,* Frag. 272, Cambridge University Press, 1957. Eudemus (350–290 BC) was a pupil of Aristotle.
10. Notice that in an infinite universe we have the dilemma of an infinite number of identical 'twins' but the basic problem still exists in a finite universe. If a finite universe is big enough, there will exist a finite number of 'twins' for each of us. The finite light travel speed means we could only ever be in contact with a finite number of our twins in an infinite universe.
11. M. Tegmark, 'Parallel Universes', *Scientific American,* May 2003.
12. The assumption that life has a zero probability of emerging by natural processes would be tantamount to ascribing it to special creation or 'intelligent design'.
13. *Quarterly Journal of the Royal Astronomical Society* 20, 37–41 (1979).
14. Ecclesiastes Iv9.
15. See P.C.W. Davies, *Nature* 273, 336 (1978) and J.D. Barrow & F.J. Tipler, *The Anthropic Cosmological Principle,* Oxford University Press, Oxford, 1986.
16. S. Webb, *Where is Everybody?,* Copernicus, New York, 2002.
17. A. Linde, 'The Self-reproducing Inflationary Universe', *Scientific American* 5, 32 (November 1994); A. Vilenkin, *Physics Letters* B 117, 25 (1982).
18. Quoted in E. Maor, *To Infinity and Beyond; a cultural history of the infinite,* Princeton University Press, 1987, p. xiii.
19. J.L. Borges, 'Avatars of the Tortoise',
20. R. Price, *Johnny Appleseed: Man and Myth,* Indiana University Press, 1954.
21. We are talking here about countable infinities in Cantor's sense. The same conclusions would hold for uncountable infinites as well.
22. C.S. Lewis, *Out of the Silent Planet* (1938), *Perclandra* (1943), *That Hideous Strength* (1945).
23. We know that certain basic altruistic actions are optimal strategies

for individuals playing evolutionary 'games' with others. This leads to cooperation of the Prisoners' Dilemma variety and behaviour, in which an 'I'll scratch your back if you'll scratch mine' philosophy is the best in the sense that any deviation from it by any player sees them worse off. However, it is interesting that human altruistic behaviour, especially that to which many people aspire and even more people admire, goes far beyond this minimal tit-for-tat altruism that arises for selfish reasons. It is only superficially altruistic.

24. J.D. Barrow and F.J. Tipler, *The Anthropic Cosmological Principle*, Oxford University Press, 1986 and J.D. Barrow, *The Constants of Nature*, Jonathan Cape, London, 2002.

chapter nine

Worlds Without End

1. I. Newton, *Opticks*, Prometheus, New York, 1952 (based on 4th edn, London, 1730), pp. 400–4.
2. C.S. Lewis, *The Lion, The Witch and The Wardrobe*, Penguin, Harmondsworth, 1976 edn, p. 49.
3. C. Bailey (ed. and trans.), *Epicurus: the Extant Remains*, Oxford University Press, 1926, p. 25.
4. Aristotle's notion of a 'world' was similar to that of the rival atomists. Steven Dick, in *Plurality of Worlds*, Cambridge University Press, 1982, p. 13, notes that Epicurus defined a 'world' as a 'circumscribed portion of the sky, containing heavenly bodies and an earth and all the heavenly phenomena'. In the same vein, Aristotle held it to denote 'the body enclosed by the outermost circumference' of space.
5. Portrait of Aristotle, copyright © akg-images.
6. Kant proposed an interesting progressive Universe in which there was a centre of high density, with matter thinning out as one moved away from the centre. Life would emerge at all distances

from the centre eventually, but its character would be very different at different distances from the centre, reflecting the local density of matter. Thus he says that:

'I am inclined to seek the most perfect classes of rational beings with more probability far away from this centre than close to it. The perfection of creatures endowed with reason, in so far as it depends on the quality of the matter in whose bonds they are confined, turns very much on the fineness of the matter which influences and determines them to their perception of the world and their reaction upon it. The inertia and resistance of matter limits very much the freedom of the spiritual being for action and the distinctness of its sense of external things; this makes its faculties dull and obtuse. So that it does not respond to their outward movements with sufficient facility. Hence, if it is supposed, as is probable, that the densest and heaviest sorts of matter are near the centre of nature, and on the contrary, that the increasing degrees in its fineness and lightness . . . are respectively at a greater distance; the result in consequence is intelligible. The rational beings, whose place of generation and abode is found nearer the centre of creation, are plunged in a rigid and immobile kind of matter which holds their powers shut up in insuperable inertness, and which of itself is just as incapable of transferring and communicating the impressions of the universe with the necessary distinctness and facility. These thinking beings will therefore have to be reckoned as belonging to the lower class. And, on the other hand, with the various distances from the universal centre, this perfection of the world of spirits, which rests on their altered dependence on matter, will grow up like an extended ladder . . . life may also go on . . . to fill all the infinitude of time and space with stages of perfection in thought that grow on to infinity, and which has to approach as it were, step by step, the goal of the supreme excellence of the Deity, without, however, being able to reach it.' *Universal Natural History and Theory of the Heavens* (1755), trans. W. Hastie (first appeared in 1900 in a volume entitled *Kant's Cosmogony,* Glasgow), University of Michigan Press, Ann Arbor, 1969, pp. 166–7. These ideas provide an interesting type of anti-Copernican argument whose significance was not recognised at the time. If the Universe has a centre and conditions vary with distance from that centre, then there will be some places where conditions make life impossible and other places where its existence is most likely. If the latter conditions were to arise near the centre, then this would provide a physical reason (rather than a philosophical prejudice) why we might find ourselves near the centre of the Universe. The modern discussion of the Anthropic Principles in cosmology has developed this insight further, see for example J.D. Barrow and F.J. Tipler, *The Anthropic Cosmological Principle,* Oxford University Press, 1986, where Kant's progressive cosmology is discussed in more detail in section 10.2.

7. I. Kant, op. cit., pp. 139–40.
8. *The Complete Works of Montaigne,* trans. D.F. Frame, Stanford University Press, 1958, p. 390.
9. In proposition 2 of his *Discovery of a World in the Moone* (1638), he writes: 'The term World may be taken in a double sense, more generally for the whole Universe, as it implies in it the elementary and aethereall bodies, the starres and the earth. Secondly, more particularly for an inferior World consisting of elements [like the Moon] . . . so that in the first sense I yeeld, that there is but one world, which is all that the arguments do prove; but understand it in the second sense, and so I affirm there may be more.'
10. See for example L. Susskind, 'The Anthropic Landscape of String Theory', http://arXiv:hepth/0302219 (2003).
11. If there are more than three dimensions, then bound structures like atoms and planets and stars cannot exist because the dimensionality determines the way the forces of Nature fall off with distance. For example, the forces of gravity and electrostatic attrac-

tion fall off like 1/(distance)$^{N-1}$ in a space of N dimensions; hence, in our 3-dimensional world we see inverse-square force laws. This link between force laws and the dimensionality of space was first noticed by Immanuel Kant.

12. In the absence of the inflationary surge the expansion is too slow to grow our whole visible part of the Universe – more than 15 billion light years across – from one of the natural quantum fluctuations.
13. When this mission was planned and launched it was called MAP, but during the data analysis phase of the project David Wilkinson, one of its guiding lights who helped pioneer observations of the microwave background radiation's temperature fluctuations back in 1967, died. When the results were announced by NASA the mission was renamed WMAP in his honour.
14. W. Allen, *Getting Even*, Random House, New York, 1988.
15. Reproduced from www.wisdomportal.com/Stanford/UniverseOrMultiverse.html by permission of Andrei Linde.
16. O. Stapledon, *Star Maker* in *Last and First Men and Star Maker*, Dover, New York, 1968, p. 419. *Star Maker* was first published in 1937.
17. See J.D. Barrow, *The Universe that Discovered Itself*, Oxford University Press, 1990, for a more detailed explanation and H.S. Leff and A.F. Rex, *Maxwell's Demon*, Inst. Phys. Bristol (2003) for a complete account of the paradox of Maxwell's demon and its resolution.
18. E.R. Harrison, 'The Natural Selection of Universes containing Intelligent Life', *Quarterly Journal of the Royal Astronomical Society* 36, 193 (1995) and 'Creation and Fitness of the Universe', *Astronomy and Geophysics*, 39, 27 (1998).
19. At first one might imagine that enormous quantities of energy would be needed to make a universe, however, small. Remarkably, the energy of a universe is *zero*: Einstein's theory of general relativity ensures that the sum of all the positive energies contributed by all the masses and other forms of energy in the universe exactly cancels out the negative energies of gravitational attraction that

exist between them all. Universes are very energy efficient.

20. Harrison calls it the 'natural selection' of universes but this is a misleading title. The process is more akin to the forced breeding of universes by unnatural selection.
21. J.D. Bernal, *The World, the Flesh and the Devil*; F. Dyson, 'Life in an Open Universe' *Rev. Mod. Phys.* 51, 447–60 (1979); J.D. Barrow and F.J. Tipler, 'Eternity is Unstable' *Nature,* 176, 453 (1978), and *The Anthropic Cosmological Principle,* Oxford University Press, 1986.
22. J.D. Barrow and S. Hervik, 'Indefinite Information Processing in Ever-Expanding Universes', *Physics Letters* B, 566, 1 (2003).
23. See J.D. Barrow, *The Constants of Nature,* Jonathan Cape, London, 2002.
24. It is not clear what chosen 'at random' means here, of course! In general, all statements of probability applied to the Universe or possible universes are inherently uncertain.
25. Paul Davies has argued very similarly that 'a general multiverse set [of all possible worlds] must contain a subset that conforms to traditional religious notions of God and design . . . But the denizens of a simulated virtual world stand in the same ontological relationship to the intelligent system that designed and created their world as human beings stand in relation to the traditional designer/creator Deity', see P.C.W. Davies, 'A Brief History of the Multiverse', *New York Times,* 12 April 2003.
26. J.K. Webb, M. Murphy, V. Flambaum, V. Dzuba, J.D. Barrow, C. Churchill, J. Prochaska, and A. Wolfe, 'Further Evidence for Cosmological Evolution of the Fine Structure Constant', *Phys. Rev. Lett.* 87, 091301 (2001).
27. I believe that the discovery that the surface area of a region of space and time limits how much information it can contain means that sub-regions of a universe cannot create perfect simulations of that universe. This means that some type of lower resolution version must be made of any simulation.
28. S. Wolfram, *A New Kind of Science,* Wolfram Inc., Champaign, IL., 2002.

29. K. Popper, *Brit. J. Phil. Sci.* 117 & 173 (1950).
30. D. Mackay, *The Clockwork Image*, IVP, London, 1974, p. 110.
31. J.D. Barrow, *Impossibility*; Oxford University Press, 1998, chapter 8.
32. Although there is a famous false argument by Herbert Simon claiming the opposite in the much cited article 'Bandwagon and Underdog Effects and the possibility of Election Predictions', *Public Opinion Quarterly* 18, 245–253 (Fall issue, 1954); it is also reprinted in S. Brams, *Paradoxes in Politics*, Free Press, New York, 1976, pp. 70–7. The fallacy arose because of the illicit use of the fixed-point theorem of Brouwer in a situation where the variables are discrete rather than continuous, see K. Aubert, 'Spurious Mathematical Modelling', *The Mathematical Intelligencer*, 6, 59, (1984) for a detailed explanation.
33. J. D. Barrow, *The Constants of Nature: From Alpha to Omega*, Jonathan Cape, London, 2002.
34. R. Hanson, 'How to Live in a Simulation', *Journal of Evolution and Technology* 7 (2001), http://www.transhumanist.com
35. David Hume, *Dialogues Concerning Natural Religion* (1779), in Thomas Hill Green and Thomas Hodge Grose (eds), *David Hume: The Philosophical Works*, London, 1886, vol. 2, pp. 412–16.
36. These questions are closely connected to the issues discussed in Ray Kurzweil's book *The Age of Spiritual Machines* (Viking, New York, 1999), concerning the appearance of spiritual and aesthetic qualities in virtual realities and forms of artificial intelligence.
37. R. Hanson, 'How to Live in a Simulation', op. cit.

chapter ten

Making Infinity Machines

1. J.F. Thomson, 'Tasks and Super-Tasks', *Analysis* 15, 1 (1954).
2. C. Wright, 'Strict Finitism', *Synthèse* 51, 248 (1982).
3. Technically a super-task means the accomplishment of a count-

ably infinite number of tasks in a finite time. If the number accomplished was uncountably infinite then we call it a hyper-task.

4. To see this call the sum of the infinite series S. Now multiply both sides of the equation $S = \frac{1}{2} + \frac{1}{4} + \frac{1}{8} + \frac{1}{16} + \frac{1}{32} + \ldots$ by $\frac{1}{2}$. So $\frac{1}{2} S = \frac{1}{4} + \frac{1}{8} + \frac{1}{16} + \frac{1}{32} + \ldots = S - \frac{1}{2}$ since the right-hand size is just the series without its first term, which is $\frac{1}{2}$. Hence, $\frac{1}{2} S = \frac{1}{2}$ and $S = 1$.
5. H. Weyl, *Philosophy of Mathematics and Natural Science,* Princeton University Press, 1949, p. 42. Weyl's mention of decision procedures and machines is interesting. Mathematics had just emerged from a pre-war period which saw the advent in print of Alan Turing's 'Turing machine', the archetypal universal computer that is indistinguishable from a human calculator (the original meaning of the word 'computer') and the question, answered in the negative by Turing, of whether a finite computing machine would be able to decide the truth or falsity of all statements of mathematics in a finite time. Turing showed that there exist uncomputable problems for which the machine would never halt if set to work on deciding them. Hence, the scenario of a calculating machine was set by these conceptions and then reinforced by the creation of actual calculating machines during the period 1939–45.
6. A Grünbaum, *Modern Science and Zeno's Paradoxes,* Allen and Unwin, London, 1968 and *Philosophical Problems of Space and Time,* 2nd edn, Reidel, Dordrecht, 1973, chap. 18. Grünbaum has completed the most detailed study of whether the Zeno journey is kinematically impossible. This is a separate issue from that of whether it is dynamically or physically possible.
7. See for example M. Black, 'Achilles and the Tortoise', *Analysis* 11, 91–101 (1950–1).
8. The processing of information requires the generation of entropy.
9. For a survey see J. Earman and J. Norton, 'Infinite Pains: The Trouble with Supertasks', in A. Morton and S. P. Stich (eds.), *Benacerraf and his Critics,* Basil Blackwell, Oxford, 1996.

10. Thomson himself thought that such devices were logical impossibilities.
11. This is the sum of the infinite geometric series that Weyl wrote down.
12. If the interval of time before the next switching became shortened to 10^{-43} sec, then the quantum gravitational structure of space and time would prevent any process occurring with the required fidelity. This quantum limit occurs after only about 148 switchings. Other physical limits would occur much sooner if the switching devices were made of atoms. A timing device of mass M that can discriminate a smallest time interval of t acts as a reliable clock for a maximum time T where $T < t^2M/h$, where h is Planck's constant. After this time, cumulative quantum fluctuations make the clock useless – see E. Wigner, 'Relativistic Invariance and Quantum Phenomena', *Rev. Mod. Phys.* 29, 255 (1957) and J.D. Barrow, 'Wigner Inequalities for a Black Hole', *Phys. Rev.* D 54, 6563–4 (1996).
13. The one-minute time limit is an illusion. We could choose to measure time in a way that shifted the one-minute end-time off to infinity and then the correspondence with the counting problem would be clear. The change from time t on $0 \rightarrow 1$ minute to a time T defined by $1/x - 1$ and which exists on the interval from $0 \rightarrow \infty$ minutes achieves this.
14. For example, the continued fraction expansion of almost every (a technical term meaning all but a collection of special cases which cease to be special cases if they are changed in an arbitrarily small way) real number will possess particular statistical properties. The continued fraction expansion of π is found to satisfy the expected statistics for as far as the expansion is followed; see J.D. Barrow, 'Chaos in Numberland', PLUS issue 11 (2000), www.plus.maths.org./issue11/features/cfractions/index.html and S. Wolfram, *A New Kind of Science*, Wolfram Inc., Champaign, IL., 2002.

15. Actually, it is likely that the decimal expansion of the number π contains all possible arithmetic sequences, including those that codify the laws of Nature. However, this is not a particularly useful piece of information, since it is necessary to extract this information from the sequence.
16. A.K. Doxiades, *Uncle Petros and Goldbach's Conjecture*, Faber, London, 1992.
17. It is interesting to note that when the list of Clay mathematics problems was drawn up a few years later, with a $1,000,000 prize for the solution of any one of them, the Goldbach conjecture did not feature among them. It appears to be an isolated problem that does not have a close link with deep structural features of mathematics. Its resolution would not unlock the solution of many other problems – or at least that is what is believed!
18. A. Wiles and R. Taylor, 'Ring-theoretic properties of certain Hecke algebras', *Ann. Math.* 141, 553–72 (1995).
19. Eli Maor, *To Infinity and Beyond: a cultural history of the infinite*, Princeton University Press, 1987, p. 33.
20. Portrait of Niels Henrik Abel reproduced from www.gap.dcs.st-and.ac.uk/~history/BiogIndex.html
21. These words of The Magnificat are used to close many parts of the liturgy in Christian religious services.
22. C.S. Chihara, 'On the Possibility of Completing an Infinite Process', *Philosophical Review* 74, 80 (1965).
23. The real numbers contain the integers as an infinite subset.
24. 2 Peter 3 v8.
25. We qualify this statement by reference to the speed in a vacuum because the speed of light in a medium will be slightly smaller than it is in a perfect vacuum. Remarkably, it is possible for objects to move in a medium with a speed exceeding the speed of light in that medium (but not exceeding the speed of light in a vacuum). When this occurs, the moving particles emit radiation. This process is called the Cerenkov Effect and is routinely observed.

It is used to detect fast-moving cosmic ray particles hitting the Earth from space.

26. There have recently been proposals by Moffat, Albrecht, Magueijo and myself that the speed of light may have changed during the very early history of the Universe. It is still possible for there to be a cosmic speed limit with the properties we have discussed. However, this would be the speed at which gravitational signals travel in a vacuum. See J. Magneijo, *Faster than the Speed of Light*, Weidenfeld, London, 2002.
27. In the case of Thomson's Lamp that we discussed earlier, there is a physical impossibility implicit in the prescription because the switch will have to have covered an infinite distance in an hour. This will require it to move faster than the speed of light and is a physical impossibility. There are ways to repair this defect in the Thomson Lamp example, see for example A. Grünbaum, 'Modern Science and Zeno's Paradoxes of Motion', in W. Salmon (ed.), *Zeno's Paradoxes*, Bobbs-Merril, Indianapolis, 1970, pp. 200–50.
28. G. Alexander, 'An Olympian Feat', *Independent*, 7 August 2004, p. 3.
29. Z. Xia, 'The Existence of Non-collision Singularities in Newtonian Systems', *Ann. Math.* 135, 411–468 (1992). The possibility of arbitrary fast expansion to infinity in infinite time is discussed in D. G. Saari and Z. Xia, 'Oscillatory and Superhyperbolic Solutions in Newtonian Systems', *J. Differential Equations* 82, 342–355 (1989).
30. J. D. Barrow, 'Sudden Future Singularities', *Classical and Quantum Gravity*, 21, L79–L82 (2004).
31. P. Cook, *Tragically, I was an only twin*, Century, London, 2002.
32. In the philosophical literature pseudo super-tasks that are not proper super-tasks are often called 'bifurcated super-tasks'.
33. I. Pitowsky, 'The Physical Church Thesis and Physical Computational Complexity', *Iyyun*, 39, 81–99 (1990).
34. An accelerated observer can experience a finite amount of proper

time along his trajectory through space and time, but there will be no point on it from which he can look back and observe an infinite elapsed history along the space and time path of any unaccelerated observer.

35. I. Pitowsky, 'The Physical Church Thesis', 81; M. Hogarth, 'Does General Relativity Allow an Observer to View an Eternity in a Finite Time?', *Foundations of Physics Letters* 5, 173–81 (1992); J. Earman and J. Norton, 'Forever is a Day: Supertasks in Pitowsky and Malament-Hogarth Spactimes', *Phil. of Science* 60, 22–42 (1993).
36. Macbeth, I. vii. I.
37. C.W. Misner, 'Mixmaster Universe', *Phys. Rev. Lett.* 22, 1071–74 (1969); C.W. Misner, K. Thorne and J.A. Wheeler, *Gravitation*, W.H. Freeman, San Francisco, 1972.
38. C.W. Misner, 'Absolute Zero of Time', *Phys. Rev.* 186, 1328 (1969).
39. This property of Mixmaster-like universes has been developed in great detail in J.D. Barrow and F.J. Tipler, *The Anthropic Cosmological Principle*, Oxford University Press, 1986.
40. J.D. Barrow and S. Hervik, 'Indefinite Information Processing in Ever-expanding Universes', *Physics Letters* B 566, 1–7 (2003). The feature used to create the indefinite processing is rather subtle. It only appears in the most general cosmological models for the future expansion of the Universe. It is the difference in the curvature of space from one direction to another that is sustained at a high enough level to generate significant temperature differences between different directions.
41. This was first pointed out by J.D. Barrow and F.J. Tipler, *The Anthropic Cosmological Principle*, Oxford University Press, 1986, p. 668.

chapter eleven

Living Forever

1. S. Ertz, *Anger in the Sky*, Harper and Bros, New York, 1943, p. 137.

2. M. de Unamuno, *The Tragic Sense of Life*, trans. J.E. Crawford Flitch, Dover, New York, 1954, p. 224.
3. Quoted in the *Observer*, 27 May 2001, p. 30.
4. An interesting discussion using game theory has been given by Steven Brams and Marc Kilgour of the different strategies and outcomes that arise in competitive circumstances where the players adopt bounded (in future time) and unbounded outlooks. When they adopt an outlook that plans into the far future, they have greater incentive to be cooperative, whereas those who take the more short-term view act less responsibly and less morally. Thus, for example, if you are engaged in terrorism or organised crime, you might believe that there would be havens from which you could not be extradited by aggrieved governments. The more short-term your outlook, the more you are likely to feel secure. But if you take a longer-term view, you would regard it as increasingly likely that the law would change or that it would be ignored; see S. Brams and D.M. Kilgour, 'Games that end in a Bang or a Whimper', in G.F.R. Ellis (ed.), *The Far-Future Universe*, Templeton Press, Radnor, PA, 2002, pp. 196–206.
5. We assume for the moment that death by natural causes does not occur, but accidental death, euthanasia and suicide may occur. In George Bernard Shaw's play *Back to Methuselah*, the Ancients did not expect to die of natural causes, but they did not expect to avoid death by 'acts of God', like earthquakes or lightning strikes; their situation differs from our own merely in the quantity of their life expectancy.
6. A. Lightman, *Einstein's Dreams*, Pantheon, New York, 1993, p. 117.
7. Ibid.
8. M. de Unamuno, op. cit., p. 248.
9. S. Butler, *Further Extracts from Notebooks*, ed. A.T. Bartholomew, Jonathan Cape, London 1934, p. 27.
10. K. Čapek, *The Makropoulos Secret* (1923), trans. P. Salver, Robert Holden, Branden, New York, 1927.
11. B. Williams, 'The Makropulos Case: reflections on the tedium of

immortality', http://www.wfu.edu/~crossaa/361/articles/bw1.htm Other modern philosophical discussions of the pros and cons of immortality have been give by T. Nagel, 'Death', in *Mortal Questions*, Cambridge University Press, 1979; F. Feldman, *Confrontations With the Reaper: A Philosophical Study of the Nature and Value of Death*, Oxford University Press, 1992, and M. Heidegger, *Being and Time*, trans. J. Maquarrie and E. Robinson, Blackwell, Oxford, 1978.

12. For a more extensive discussion of whether science will ever complete its account of the Universe, and what such a statement might mean, see J.D. Barrow, *Impossibility*, Oxford University Press and Vintage, 1998.
13. A countably infinite number of future moments, or thoughts, would make no impression on an uncountably infinite universe of potential information and experience.
14. A.C. Clarke, *The City and the Stars*, Harcourt, Brace and World, New York, 1956.
15. S.L. Clark, *How to Live Forever: science fiction and philosophy*, Routledge, London, 1995, p. 16.
16. Of course, the Christian doctrine is that our human nature will be changed, and perfected, fitting us for eternal existence. The exploration of radically different characteristics which might mitigate the simple problems of everlasting life have been treated most imaginatively by science fiction writers. A fascinating survey of the most interesting philosophical ideas about living forever that are to be found in this literature has been given by Stephen Clark in his book *How to Live Forever: science fiction and philosophy*, op. cit. Here is his account (p. 12) of Gulliver's encounter with immortals in *Gulliver's Travels* by Jonathan Swift: 'Gulliver's third voyage was to the islands of Laputa, Balnibarbi and the rest. On one such island he learns of the Struldbruggs, rare mutations who are doomed to "live forever". He is exhilarated by the thought, and fantasises how he would himself behave if he had been so fortunate. But Struldbruggs age, acquire all the faults of age and are denied (for the sake of

their mortal descendants) any of its privileges. They are the least fortunate of human beings, destined to outlive their times, their friends, the very language of their youth, and subject to all the little pains and humiliations of age. If they were allowed to own any property they would accumulate it greedily, past any sensible use of it . . . [But] would living forever be essentially a bad idea, even if all ordinary advantages of health and good sense were added?

'A much more serious problem . . . is that a solitary immortal will always be losing friends, homes, civilizations. Swift's Struldbruggs cannot even communicate with later-born immortals, since their mother tongues have changed so drastically, and they have lost any capacity to learn a new one . . . What of friends and families, and familiar worlds? Maybe immortals must keep company especially with each other: mortal mayflies could not hold their attention long. Or would immortals be the worst company of all, with thousands of years to find each other's habits more exasperating? . . . [they may discover] they have nothing in common but their immortality. Would they cultivate mere mortals to occupy familiar slots, surrounding themselves with good examples of a type they once knew well? Would they notice which individuals fulfilled those roles, or would they care? . . . another serious problem emerges: boredom. What can occupy this endless, hurrying time, which must be managed to seem familiar even at the cost of being stiflingly the same? Take another set of . . . immortals . . . only a tiny percentage of these immortals can find anything of such interest to do more than "kill time" . . . they are incapable of living coherently or sanely. They have done everything so many times before that nothing is worth doing.'

17. It is interesting that the Buddhist tradition of *nirvana* is a state in which all desires are extinguished and hence eternal existence is imagined to have stable psychological features.
18. Song written by Marvin Hamlisch, Alan Bergman and Marilyn Bergman, from the film *The Way We Were* (1973); see http://lyrics playground.com/alpha/songs/t/thewaywewere.html

19. The infinite series $1 + \frac{1}{2} + \frac{1}{3} + \frac{1}{4} + \frac{1}{5} + \frac{1}{6} + \frac{1}{7} + \frac{1}{8} + \ldots$ has an infinite sum.
20. C.S. Lewis, *Perelandra* (1943), Pan, London, 1953, p. 200.
21. M. Dibdin, *Medusa,* Faber, London, 2003, p. 248.
22. For a fuller discussion see J.D. Barrow, *Impossibility,* Vintage, London, 1998.
23. See D. Deutsch, 'Quantum Mechanics Near Closed Time-like Lines', *Phys. Rev.* D, 44, 3197 (1991); D. Deutsch and M. Lockwood, 'The Quantum Physics of Time Travel', *Scientific American,* 270, March 1994, 68–74; and D. Deutsch, *The Fabric of Reality,* Penguin, London, 1997.
24. M. MacBeath, 'Who was Dr Who's Father?' *Synthèse* 51, 397–430 (1982); G. Nerlich, 'Can Time be Finite?', *Pacific Philosophical Quarterly* 62, 227–39 (1981).
25. W. Churchill, speaking of the Government in the House of Commons, *Hansard,* 12 November 1936, col. 1107.
26. S.W. Hawking, 'The Chronology Protection Hypothesis', *Phys. Rev.* D 46, 603 (1992); M. Visser, *Lorentzian Wormholes – from Einstein to Hawking,* Amer. Inst. Phys., New York 1995.
27. A. Huxley, 'Wordsworth in the Tropics', *Do What You Will,* Chatto & Windus, London, 1929.
28. S. Butler, *Erewhon Revisited,* 1901, chap. 14.
29. R. Silverberg, *Up the Line,* Ballantine, New York, 1969.
30. J. Varley, *Millennium,* Berkley, New York, 1983.
31. See chapter 9 of J.D. Barrow and F.J. Tipler, *The Anthropic Cosmological Principle,* Oxford University Press, Oxford, 1986.
32. H.A. Dobson, 'Paradoxes of Time', *Proverbs in Porcelain,* Kegan Paul, London, 1905.
33. M.R. Reinganum, 'Is Time Travel Impossible?: A Financial Proof', *Journal of Portfolio Management* 13, 10–12 (1986).
34. D. Adams, *The Restaurant at the End of the Universe,* Tor Books, London, 1980.
35. H. Hupfeld, *As Time Goes By,* song (1931).

36. L. Dwyer, 'Time travel and changing the past', *Philosophical Stud.*, 27, 341–50 (1975); see also 'Time travel and some alleged logical asymmetries between past and future', *Can. J. of Phil.* 8, 15–38 (1978); 'How to affect, but not change, the past', *Southern. J. of* Phil. 15, 383–5 (1977).

37. The American philosopher, David Malament, discusses the common view that, because of Grandmother Paradoxes, 'time-travel . . . is simply absurd and leads to logical contradictions. You know how the argument goes. If time travel were possible, one could go backward in time and undo the past. One could bring it about that both conditions P and not-P obtain at some point in spacetime. For example, I could go back and kill my earlier infant self, making it impossible for that earlier self ever to grow up to be me. I simply want to remark that arguments of this type have never seemed convincing to me . . . The problem with these arguments is that they simply do not establish what they are supposed to. To be sure, if I could go back and kill my infant self, some sort of contradiction would arise. But the only conclusion to draw from this is that if I tried to go back and kill my infant self then, for some reason, I would fail. Perhaps I would trip at the last minute, The usual arguments do not establish that time travel is impossible, but only that if it were possible, certain actions could not be performed.' (*Proc. Phil. Science Assocn.*, 2, 91 (1984)) Another distinguished philosopher who swam against the tide, and argued for the rationality of time travel in the face of the Grandmother Paradoxes, was David Lewis. In 1976, he wrote in his review 'The Paradoxes of Time Travel', *Amer. Phil. Quarterly* 13, 142–152 (1976) that 'Time travel, I maintain, is possible. The paradoxes of time travel are oddities, not impossibilities. They prove only this much, which few would have doubted: that a possible world where time travel took place would be a most strange world, different in fundamental ways from the world we think is ours.'

38. F. Nietzsche, *Joyful Wisdom*, F. Ungar Publ. Co., 1964, pp. 270–1.
39. On first meeting his political rival, Senator John McCain, in 2000.
40. The legendary image of an army of monkeys typing letters at random and eventually producing the works of Shakespeare seems to have gradually been constructed over a long period of time. Jonathan Swift's *Gulliver's Travels* (1782) tells of a professor of the Grand Academy of Lagado who aims to generate a catalogue of all scientific knowledge by having his students continuously generate random strings of letters by means of a mechanical printing device. The first mechanical typewriter was patented in 1714. The monkeys turn up in 1909 in a version of the scenario written by the French mathematician Emile Borel in his book on probability, *Élements de la théorie des probabilités* (Paris, 1909), where he suggests that the randomly typing monkeys would eventually produce every book in France's Bibliothèque Nationale. Arthur Eddington takes up the analogy in his book *The Nature of the Physical World* (Cambridge University Press, 1928), where he changes the library (p. 72): 'If I let my fingers wander idly over the keys of a typewriter it *might* happen that my screed made an intelligible sentence. If an army of monkeys were strumming on typewriters they *might* write all the books in the British Museum.' Eventually this oft-repeated example homed in on the works of Shakespeare as the candidate for random re-creation. In fact, there is now a website which simulates an ongoing random striking of typewriter keys and then does pattern matches against the complete works of Shakespeare to identify matching character strings. This simulation of the monkeys' actions began on 1 July 2003 with 100 monkeys and the population of monkeys has doubled every few days ever since. They have produced more than 10^{35} pages, each requiring 2000 keystrokes. So there have been many matches of 19 letters (3 words) and at present the record is 21. To check how the monkeys are doing see http://user.tninet.se/~ecf599g/aardasnails/java/Monkey/webpages/. A running and daily record

is kept of record strings. For example today I see that one of the 18-character strings that the monkeys have generated contains the snatch:

. . . Theseus. Now faire UWfIlaNWSK2d6L;wb . . .

The first 18 characters match part of an extract from *A Midsummer Night's Dream* that reads

. . . us. Now faire Hippolita, our nuptiall houre . . .

The 21-character record is

. . . KING. Let fame, that wtIA"'yh!"VYONOvwsFOsbhzkLH . . .

which matches 21 letters from *Love's Labour's Lost*

KING. Let fame, that all hunt after in their lives, Live regist'red upon our brazen tombs,
And then grace us in the disgrace of death; . . .

Which all goes to show it's just a matter of time!

Index

Numbers in italics refer to Figures.